Round Table Conference on the CRYOGENIC PRESERVATION OF CELL CULTURES

U.S. National Committee for the
International Institute of Refrigeration

NATIONAL ACADEMY OF SCIENCES

Summaries of Presentations

ROUND TABLE CONFERENCE

ON

THE CRYOGENIC PRESERVATION OF CELL CULTURES

Edited by
A. P. Rinfret, Ph.D.
B. LaSalle, L.Ag.Sc., D.V.M.

A meeting held under the auspices of the United States National Committee for the International Institute of Refrigeration of the National Research Council, at the National Academy of Sciences, Washington, D.C.

QH324.9
C7
A1
R68
1975

National Academy of Sciences
Washington, D.C.
1975

The National Research Council, under the cognizance of the National Academy of Sciences and the National Academy of Engineering, performs study, evaluation, or advisory functions through groups of individuals selected from academic, governmental, and industrial sources for their competence and interest in the subject under consideration. Members serve as individuals contributing their personal knowledge and judgments and not as representatives of any organization in which they are employed or with which they might be associated.

The project which is the subject of this report was approved by the Governing Board of the National Research Council, acting in behalf of the National Academy of Sciences. Such approval reflects the Board's judgment that the project if of national importance and appropriate with respect to both the purposes and resources of the National Research Council. The opinions expressed by the authors of papers presented at this conference do not necessarily reflect the views of the National Academy of Sciences and are the sole responsibility of the authors.

Library of Congress Catalog Card No. 75-7690
International Standard Book No. 0-309-02344-0

Available from:

Printing and Publishing Office
National Academy of Sciences
2101 Constitution Avenue, N.W.
Washington, D.C. 20418

Printed in the United States of America

TABLE OF CONTENTS

Conference Participants ... iii

Background to the Conference ... vi
   A. P. Rinfret
   Vice President, Commission C-1
   International Institute of Refrigeration

Introduction ... viii
   Bernard La Salle, L.Ag.Sc., D.V.M.
   Staff Veterinarian
   USDA, APHIS Biologics Licensing & Standards

The Role of Liquid Nitrogen Refrigeration at the
American Type Culture Collection ... 1
   John E. Shannon, Robert L. Gherna, and Shung C. Jong
   American Type Culture Collection (ATCC)

Some Considerations in the Cryogenic Preservation
of Cultures ... 9
   William H. Trejo
   Department of Microbiology
   The Squibb Institute for Medical Research

Preservation and Shipping of Industrial Starter Cultures ... 16
   Robert L. Sellars
   Chr. Hansen's Laboratory, Inc.

Nitrogen Preservation of Stock Cultures of Unicellular
and Filamentous Microorganisms ... 22
   Alma Dietz
   Infectious Diseases Research
   The Upjohn Company

Preparation, Storage and Recovery of Free-Living,
Nonencysting Ciliated Protozoa ... 37
   Ellen M. Simon
   University of Illinois
      and
   Miriam Flacks
   University of California at Santa Barbara

Preservation of Cultures in Routine Microbiological Assay   50
Operations
   Walter T. Sokolski
   Infectious Diseases Research
   The Upjohn Company

Cryopreservation of Protozoa and Helminths   60
   Kenneth W. Walls
   Center for Disease Control
   U.S. Dept. of Health, Education, and Welfare

Comments on Various Aspects of the Cryogenic Preservation   71
of Cell Cultures
   A. P. Rinfret
   Corporate Research Laboratory
   Union Carbide Corporation

CONFERENCE PARTICIPANTS

Dr. D. E. Baldwin
Director, Bio Production
Fort Dodge Laboratories
Fort Dodge, Iowa 50501

Dr. Jack Barry
Assembly of Life Sciences
Division of Medical Sciences
National Academy of Sciences
2101 Constitution Avenue
Washington, D. C. 20418

Dr. Charles H. Campbell
Research Leader, Cytology
  Investigations
Plum Island Animal Disease Center
ARS, U.S. Department of Agriculture
Post Office Box 848
Greenport, Long Island,
  New York 11944

Dr. Z. F. Chmierewicz
Associate Prof. of Biochemical
  Pharmacology
State University of New York at
  Buffalo
Buffalo, New York 14214

Dr. Manuel Coria
National Animal Disease Center
North Central Region
Agricultural Research Service
U.S. Department of Agriculture
Post Office Box 70
Ames, Iowa 50010

Mr. Harlan E. DenBeste
Vice President
Customer Services
Grand Island Biological Co.
3175 Staley Road
Grand Island, New York 14072

Miss Alma Dietz
Infectious Diseases Research
The Upjohn Company
Kalamazoo, Michigan 49001

Dr. Emil F. Gelenczei
Vice President & General Manager
Agri-Bio Corporation
P. O. Box 280
Ithaca, New York 14850

Dr. R. R. Grunert
Research Associate
Microbiology Section
Stine Laboratory
E. I. duPont de Nemours & Co.,
  Inc.
Newark, Delaware 19711

Dr. William C. Haynes
Principal Bacteriologist
ARS Culture Collection Research
Fermentation Laboratory
U.S. Department of Agriculture
1815 N. University
Peoria, Illinois 61604

Dr. Werner P. Heuschele
Jensen-Salsbery Laboratories
Div. Richardson-Merrell, Inc.
2000 South 11th Street
Kansas City, Kansas 66103

Dr. James A. House, Manager
Biological Research
Pitman-Moore, Inc.
P. O. Box 344
Washington Crossing, N. J. 08560

Dr. Julian Kramer, Director
Division of Drug Biology
Food and Drug Administration
Room 2831: Std. Adm. Code HFD-410
200 C Street, S.W., Building FB-8
Washington, D. C. 20204

Dr. Bernard LaSalle, Staff
  Veterinarian
APHIS, Biologics Licensing &
  Standards
U.S. Department of Agriculture
Room 831, Federal Building No. 1
Hyattsville, Maryland 20782

Dr. Hiram N. Lasher
Sterwin Laboratories, Inc.
P. O. Box 537
Millsboro, Delaware 19966

Dr. Alan P. MacKenzie
American Foundation for
  Biological Research
RFD 5, Box 137
Madison, Wisconsin 53704

Dr. Shraga Makover
Department of Microbiology
Hoffmann-La Roche, Inc.
Nutley, New Jersey 07110

Dr. William McFarland
Veteran's Hospital
50 Irving Street, N.W.
Washington, D. C. 20422

Dr. Jack McGonigle
Research Director
Burns-Biotec Laboratories, Inc.
9456 Marshall Drive
Lenexa, Kansas 66215

Dr. Busch Meredith, Vice President
Syntex Agribusiness, Inc.
Animal Health Division
P. O. Box 863
Des Moines, Iowa 50304

Dr. Robert A. Oppermann
Senior Microbiologist
Corning Glass Works
Corning, New York 14830

Dr. Gerald Orgel
Air Products & Chemicals, Inc.
P. O. Box 538
Allentown, Pennsylvania 18105

Mr. W. T. Pentzer
13216 Ingleside Drive
Beltsville, Maryland 20705

Mr. A. R. Ramee, Jr.
Product Manager
Linde Division
Union Carbide Corporation
270 Park Avenue, 8th Floor
New York, New York 10017

Dr. Arthur P. Rinfret
Union Carbide Corporation
Tarrytown Technical Center
Old Saw Mill River Road
Tarrytown, New York 10591

Dr. Robert L. Sellars
Chr. Hansen's Laboratory, Inc.
5015 West Maple Street
Milwarkee, Wisconsin 53214

Dr. John Shannon
American Type Culture Collection
  (ATCC)
12301 Parklawn Drive
Rockville, Maryland 20852

Mr. Harvey E. Sheppard, Staff
  Officer
International Union of Biological
  Sciences
National Academy of Sciences
2101 Constitution Avenue
Washington, D. C. 20418

Dr. Ellen M. Simon
Department of Zoology
University of Illinois
Urbana, Illinois 61801

Dr. Walter T. Sokolski
Infectious Diseases Research
The Upjohn Company
Kalamazoo, Michigan 49001

Dr. Russell B. Stevens
Executive Secretary
Division of Biological Sciences
National Academy of Sciences
2101 Constitution Avenue
Washington, D. C. 20418

Dr. William Taylor
National Cancer Institute
Building 37, Room 4D18
Bethesda, Maryland 20014

Mr. Larry Thompson
Philips Roxane, Inc.
2621 North Belt Highway
St. Joseph, Missouri 64502

Dr. William H. Trejo
Department of Microbiology
The Squibb Institute for Medical
  Research
P. O. Box 4000
Princeton, New Jersey 08540

Mr. H. T. Tung
Manager, Product Development
Vineland Laboratories, Inc.
2285 East Landis Avenue
Vineland, New Jersey 08360

Mr. Monroe Vincent
Microbiological Associates, Inc.
4813 Bethesda Avenue
Bethesda, Maryland 20014

Dr. Kenneth W. Walls
Parasitology Branch
Bureau of Laboratories
Center for Disease Control
Atlanta, Georgia 30333

Dr. Robert H. Yager
Executive Secretary
Inst. of Laboratory Animal Resources
National Academy of Sciences
2101 Constitution Avenue
Washington, D. C. 20418

Mr. Arthur J. Yaillen
Production Director
Fromm Laboratories, Inc.
Grafton, Wisconsin 53024

NAS Staff:

Dr. Nelson T. Grisamore
Staff Executive, USNCIIR

Miss Joan M. Sieber
Administrative Secretary

BACKGROUND TO THE CONFERENCE

One of the functions of the U.S. National Committee for the International Institute of Refrigeration (USNCIIR) in association with Commissions of the IIR is to disseminate the results of scientific work in those fields covered by the Commissions. In association with Commission C-1, Biology and Medicine, of the IIR the USNCIIR presents to representatives of industry, the academic community, and government agencies the results of scientific work in fields in which low temperatures and biological systems interact. This conference on the cryogenic preservation of cell cultures is the second held at the National Academy of Sciences. The first, held in 1972, dealt with a broader range of subject matter including the use of freezing temperatures in surgery, the preservation of blood in the frozen state, problems in the preservation of organs, evidence for the fertility of human spermatozoa preserved cryogenically, and other topics of current interest.

To encourage discussion between scientists presenting their work and those attending the conference, the sessions are held in round table form, numbers are limited, and those invited are believed to have specific interest in the topics being considered. The proceedings are then summarized and made available on request to the National Academy of Sciences (USNCIIR). The section of commentary I have prepared deals with points which were discussed beyond the actual texts of the presentations or which deserve special consideration by anyone contemplating the establishment of a bank of frozen cells as a new undertaking. Apart from its function as a record of the proceedings of the conference, it is hoped that this publication will have practical reference value to those confronted with operational problems which the authors have obviously solved.

Having on earlier occasions heard the present authors speak of their work or visited their laboratories, it was not difficult to decide that others would welcome the opportunity to participate in discussion with them at greater depth than is ordinarily afforded at large meetings. To their willingness to contribute to this conference the National Committee and I are deeply indebted. Among the informal contributors to the conference it is a pleasure to acknowledge Alan Mackenzie who brought to bear in his verbal comments the special view of the biophysicist.

To achieve its objectives a conference such as this requires a sustained organizational effort. I am particularly grateful to Nelson Grisamore and Joan Sieber of the National Committee secretariat, to W. T. Pentzer of the International Institute of Refrigeration, and to Bernard LaSalle of the Department of Agriculture who not only participated in the organization of the conference but contributed to its content as well.

A third conference is planned by the United States National Committee for 1975. It will deal with the science and technology of biological preservation over a temperature range extending from freezing point of water to about -100 C. Any who have special interest in presenting their work or participating as auditor and conferee may advise the secretariat of the National Committee.

Tarrytown, N. Y.
September, 1974

Arthur P. Rinfret
Vice President
Commission - C-1
International Institute of Refrigeration

INTRODUCTION
CRYOGENIC PRESERVATION OF LIVE CELLS-IMPORTANCE

We owe the cryogenic preservation of live bacterial and animal cells to continuing scientific advances which may be considered as the least heralded in regard to their merits and benefits.

Through cryogenic preservation, it may be possible to indefinitely stop biological time, stop aging, halt all the life processes of bacterial and animal cells, and return them to full life activities at any point in time.

Complete plants have been produced from single cells. If this can be done with animal cells, it may be possible to produce identical tissues and organs in unlimited numbers. It is refreshing to think that through applied cryobiology the world may be improved.

The cryogenic preservation of live cells has made possible the following accomplishments:

1. Establishment of repositories for worldwide distribution of microorganisms, viruses, plant and animal cells, semen, blood, tissues, protozoa, and metazoa.

2. Economy of time and material required for continuous passage of cells.

3. Elimination of the genetic variations due to continuous passages.

4. Continuous availability for an unlimited number of years, of unlimited supplies of fully characterized and tested cells, of the same origin, the same passage, with known purity, properties, and potential.

5. Unlimited production of consistent high quality bacterial and viral vaccines, toxoids, antitoxins, diagnostics, starter cultures, enzymes, research products, reagents, national, and international standards.

6. Accumulation of a worldwide wealth of data on cellular stability, previously inconceivable.

7. Establishment, production, and distribution of patented strains.

The preservation of live cells at very low temperatures has been achieved through the application of cryogenics and cryobiology. Cryogenics, as applied to biomedical situations, has been simply defined as the study and use of extreme cold, between -100 C and -273 C. Minus 100 C is colder than the coldest natural temperature ever recorded on earth.

The cryogenic storage temperatures now commonly used for live cells are those of liquid nitrogen (-196 C) and liquid nitrogen vapor for higher storage temperature.

Other cryogenic liquefied gases available for research and application are:

    Oxygen O_2 -183 C
    Argon   Ar -186 C
    Neon   Ne -246 C
    Hydrogen H_2 -253 C
    Helium   He -269 C

Cryogenic research and applications include:

    Transport refrigeration
    Electronics
    Rubber and plastics
    Superconductivity
    Metallurgy
    Nuclear physics
    Cryobiology

Cryobiology research and applications include:

    Freezing and storage of live cells and biological materials
    Preservation of vaccines and diagnostics
    Preservation of enzymes
    Preservation of blood and components
    Preservation of semen for artificial breeding
    Preservation of cheese and yeast cultures
    Preservation of multicellular organisms
    Preservation of tissues and organs
    Cryosurgery

It has been rightfully said that no single person can write with authority on all aspects of cryobiology.[1] However, I will mention certain fundamentals likely to be applicable to much that will be presented in the following text.

---

[1] H. T. Meryman *et al*, 1966, Cryobiology, 775 pp. Academic Press.

The production of cryogenic temperatures is a relatively simple undertaking, as compared to the determination of the effects of these temperatures on biological systems such as bacteria, protozoa, plants, animals, live cells, viruses, their components and environments.

The complexity of live cells, their components, and their dynamic relationships makes difficult the analysis of change brought about by low temperature in such systems.

The temperature range compatible with the life of most animals and plants is 5 to 45 C. Above 45 C proteins and nucleic acids are denatured or inactivated.

A temperature fall of 10 C usually results in a 50% rate reduction of most biochemical and physiological functions. Temperatures just above 0 C may deeply alter cellular metabolism and functions.

At freezing and sub-freezing temperatures many changes can occur such as formation of ice crystals, exosmosis, increased solubility of gases, dehydration, increased concentration of electrolytes, colloids, salts, carbohydrates, lipids, proteins, lower pH, changes in heat and electric conductivity, decreased activity of some enzymes (dehydrogenase), increased activity of other enzymes (catalase, lipase), accumulation of intermediary products, reduced intermolecular spaces, increased molecular contacts, disruption of weak hydrogen bonds, breakdown of emulsions, formation of gels, precipitates, abnormal bonds, cross linkages, polymers, folding and distortion of large molecules, loss of cell membrane integrity, cellular invasion by toxic and mutagenic salts, nucleic acid eutectic concentration of all solutions, solidification, immobilization of all molecules, and drastic changes of physical, chemical, and biological properties.

Practically all the undesirable effects of freezing and thawing may be prevented or minimized to a negligible level for the cryogenic preservation of live cells through a few simple basic steps.

Fresh cultures of cells in fresh growth medium are used.

The cell count is adjusted to 2 to 6×10^6 cells per ml. Glycerol or dimethyl sulfoxide (DMSO) are added and mixed immediately to final concentrations of 5 to 15%.

The suspension is filled in cryogenic ampules and held for 30 minutes at room temperature, to allow the establishment of osmotic equilibrium between the cells and cryoprotective medium.

The temperature is lowered at the rate of 1 to 3 degrees per minute to -30 C followed by a more rapid rate of 15 to 30 degrees per minute to -100 C or lower, and transfer to -196 C liquid nitrogen.

Through this simple procedure the medium is frozen slowly, drawing water from the cells until they reach a eutectic point and solidify into a state of reduced molecular motion.

These cells may be returned to active life by placing a frozen ampule directly from liquid nitrogen storage into a water bath at 37 C for rapid thawing followed immediately by the slow addition of culture medium to allow slow readjustment of cellular osmotic pressure and cryopreservative dilution without physical or chemical damage.

This process is extensively used throughout the world. The American Type Culture Collection has reported thawing cells with 95% viability after 10 years of cryogenic storage.

Cryobiology also has an important role in veterinary medicine, a role of particular interest to me. In fiscal year ending June 30, 1973, 15.6 billion doses of Veterinary Biologics were produced under USDA licenses.

Of these, 2.2 billion doses were produced with cell cultures, many of which were started from master seeds kept in liquid nitrogen.

Approximately 1.5 billion doses of Marek's Disease Vaccine were marketed in liquid nitrogen as live turkey herpesvirus in live chicken cells.

Our distinguished contributors, in the text which follows, will tell us much more about the cryogenic preservation of cells and its important applications.

Hyattsville, Maryland
June, 1974

Bernard LaSalle, L.Ag. Sc., D.V.M.
Staff Veterinarian,
USDA, APHIS Biologics Licensing
    and Standards

THE ROLE OF LIQUID NITROGEN REFRIGERATION
AT THE
AMERICAN TYPE CULTURE COLLECTION

John E. Shannon, Robert L. Gherna, and Shung C. Jong
American Type Culture Collection

The American Type Culture Collection (ATCC), as one of the most comprehensive repositories of standard microbiological reference cultures in the world, obviously has to concern itself with the development and utilization of effective methods for the long-term preservation of microbial cultures in a stable condition. The chief methods employed by the ATCC are preservation in the freeze-dried state at +4 C or -65 C via mechanical refrigeration and preservation in the frozen state at -196 C via liquid nitrogen refrigeration. Preservation in the frozen state at -65 C in mechanical refrigerators is also employed for a number of "wet" virus preparations.

Prior to 1960 over 95% of the bacteria at the ATCC and approximately 60% of the fungi (sporulating forms) were preserved exclusively in the freeze-dried state at +4 C. In 1960 the ATCC and collaborators initiated a developmental research program to explore the effectiveness of preservation in the frozen state at liquid nitrogen temperatures as an alternative method for the long-term storage of a variety of forms. Research on the preservation of animal cells in liquid nitrogen[1,2,3] led to the establishment of a bank of animal cell cultures at the ATCC[4,5] and effective procedures were also developed at the ATCC for the preservation of nonsporulating fungi[6,7] plant rusts[8,9], bacteria[10,11], bacteriophages[12,13], mycoplasmas[14], protozoa[15], and algae[16]. In the course of the above efforts the ATCC cooperated closely with advisors and with industrial organizations such as the Linde Corporation, Cryenco, Canalco, and others in the development and field testing of liquid nitrogen refrigerators, programmed freezing units, and associated equipment.

The approximate numbers of strains maintained at the ATCC and the number of these strains that are preserved in liquid nitrogen are shown in Table I. The Collection of Animal Cell Lines currently banks over 320 lines (derived from approximately 40 different species). All are stored in liquid nitrogen. In the Collection of Fungi approximately 4,500 strains of nonsporulating forms, which could not be preserved by freeze-drying, are now stored in liquid nitrogen. The majority of these strains are fungi imperfecti and ascomycetes; the balance are basidiomycetes, phycomycetes, and slime molds. Although almost all of the strains in the Collection of Bacteria can be preserved in the freeze-dried state, storage in the frozen state at liquid nitrogen temperatures is now used as a back-up system for cultures of all of the Type and

TABLE I

NUMBER OF STRAINS PRESERVED IN LIQUID NITROGEN AT THE ATCC
RELATIVE TO THE TOTAL STRAINS IN EACH COLLECTION

	Total Strains	Strains Preserved in Liquid Nitrogen
Animal Cell Lines	320	All
Animal Viruses and Rickettsiae	730	500
Bacteria	10,000	1,200
Fungi	7,500	4,500
Plant Viruses	220	All
Protozoa and Algae	350	260

Neotype strains (approximately 1200). For bacteria we have also found liquid nitrogen storage in the frozen state to be useful in the following cases: Some bacteria such as the thermophilic actinomycetes and *Treponema pallidum* strain Kazan 8 either suffer a drop in viability during the freeze-drying procedure, or have a short shelf-life during storage at 5 C. There are also some bacteria that do not survive the lyophilization process at all. A few examples of this group of microorganisms are the bacteria *Spirillum volutans*, *Acholeplasma bactoclastica*, certain L-phase variants, and blue-green bacteria. All of the above strains can be successfully recovered from storage in the frozen state with very little loss of viability. With preservation of bacteria by routine freeze-drying methods there is usually a loss in viability of one or two logs or more. However, in the course of the routine preservation of many bacteria in the frozen state at liquid nitrogen temperatures we often find that the loss in viability, as determined by colony forming ability, is less than one log. The greater efficiency in recovery of the cells from the frozen state is important because it minimizes the possibility of the selection of cells resistant to the technical manipulations inherent in the method of preservation employed. A third group of bacteria can be freeze-dried with good recovery but lose certain physiological and/or morphological characteristics after lyophilization: for example, some strains of bacteria such as *Clostridium botulinum* type E lose their ability to produce toxin, a few strains of *Salmonella* lose their somatic antigens, and some antibiotic producers show a diminished yield of antibiotic synthesis after the freeze-drying process. The above properties are retained, however, when such strains are preserved in the frozen state at liquid nitrogen temperatures. We have had similar experiences with fungi and other forms. Certain cell-associated viruses such as varicella and cytomegaloviruses are stored in liquid nitrogen because they are more stable than when they are stored at -65 C. Storage in the vapor phase of liquid nitrogen is also used as a backup system for most of the ATCC's animal and plant virus seed stocks (Table I). The majority of the protozoa and algae at the ATCC (Table I) are also preserved in liquid nitrogen. All of the above-mentioned strains are described or referenced in a series of catalogues available from the ATCC[5,17,18,19].

Although glycerol and dimethyl sulfoxide (DMSO) appear to be almost equally effective in preserving many types of microorganisms and animal cells, glycerol is best for certain strains and DMSO for others. In our experience, glycerol or DMSO at the relatively low final concentration of 5% (v/v) in fresh culture medium is quite adequate for the preservation of most animal cells and bacteria. For the fungi, however, we usually use 10% glycerol in distilled water. Use of the cryoprotective agent at the low (5%) level has the advantage that the removal of the agent is facilitated upon dilution with fresh culture medium at the time of thawing.

We recommend that the cryoprotective agent employed be of reagent grade. Also, because of the accumulation of oxidative breakdown products we avoid using an opened bottle for too long a period of time. To minimize this problem the glycerol and DMSO are sterilized in units of small volume that are used only once. The volume of glycerol required for a particular freeze is sterilized by autoclaving for 15 minutes at slow exhaust. The DMSO is filtered, using 03 porosity Selas filter candles, collected in 10-15 ml quantities in test tubes, and then stored, frozen, at +5 C (DMSO freezes at +18 C).

The standardized freezing procedures we employ are essentially those worked out for animal cells[5,20] and fungi[6,7] which were adapted, with appropriate variations, to the routine freezing of all microorganisms and animal cells at the ATCC. Most of the details of the basic procedures, as given below, are described in Chapter 8 of Tissue Culture Methods and Applications edited by Paul F. Kruse and M. K. Patterson, Jr., 1973, Academic Press, N.Y. Differences in the handling of bacteria and fungi will be pointed out below following the description of our general procedure for the freezing of animal cells: Twenty-four hours after a fluid renewal the cells are harvested in the appropriate culture medium to which 5% or 10% glycerol or DMSO has been added as a cryoprotective agent. One ml aliquots containing approximately 2 to 6 million cells are dispensed into 1.2 ml ampules. It is important during the above manipulations of the cells to avoid alkaline drift of the culture medium by gassing with 5 or 10% CO_2 in air, as necessary. The ampules are then sealed with a pull-seal to minimize formation of pinhole leaks. After sealing, the ampules are allowed to stand for approximately 30 minutes at +5 C in a 0.05% methylene blue solution. This allows time for equilibration of the cryoprotective agent and also for the dye to leak into improperly sealed ampules. The cells are then frozen in a freezing unit at a programmed rate with a 1 to 2 degree drop in temperature per minute to -30 C, then a more rapid drop of 15 to 30 degrees per minute to -150 C. After this the ampules are immediately transferred to storage in liquid nitrogen at -196 C.

We use heavy-walled borosilicate (bull-semen) ampules[i] of 1.2 ml capacity in most instances. Special wide-mouth ampules[i], however, are used for the fungi. No washing of the ampules is necessary. We label the ampules by means of a Markem Labeling Machine (Model 135A)[ii] using a specially formulated ink (Markem #7224K) that is resistant to ultra-low temperatures and brief exposure to ordinary laboratory solvents such as alcohol. With this machine one can conveniently and rapidly mark small (one ml), large (10 ml) ampules, or even bottles up to about 4 inches in diameter. After marking, the ampules are put in an aluminum ampule rack and are placed in a hot-air oven for 60 minutes at 120 C to anneal the ink. Upon cooling, specially fabricated glass ampule caps are placed over the neck of each ampule in the rack, and then the ampules are sterilized by dry-heat sterilization (270 C for two hours). If it is necessary to label the ampules by hand this may be done by using an ordinary stick pen and most laboratory inks, provided the ink is subsequently annealed.

For the animal cells we use a specially designed apparatus to dispense uniform aliquots into the ampules. The apparatus consists of a Wheaton Celstir unit[i] modified with a side arm near the base for attaching a Cornwell pipetting unit. The unit contains a teflon-coated magnetic stirring bar for gentle agitation of the cell suspension, and is also equipped with side arms near the top for continuous gassing with 5-10% CO_2 in air mixture to maintain the proper pH. One ml aliquots of the cell suspension are dispensed into the ampules by means of the Cornwall syringe. In the case of bacteria, the cells are usually harvested at the beginning of the stationary phase. Sporulating fungi are harvested after the formation of mature spores. The nonspore forming fungi are harvested by cutting hyphal tips from agar plates with a sterile cork borer. Sometimes a blended suspension of hyphal fragments is used. The agar plugs containing the hyphal tips or the bacterial cells or fungal spores are then suspended in the appropriate cryoprotective mixture. A half milliliter of the suspension or several agar discs are transferred to cotton-plugged, sterile 1.2 ml ampules. The ampules are precooled to +5 C in order to prevent overheating the sample during the sealing operation and are then attached to prelabelled canes[iii] prior to freezing.

The proper sealing of ampules and testing for leaks is a must if the ampules are to be stored completely immersed in the liquid phase of a liquid nitrogen refrigerator. We seal the ampules using a Kahlenberg-

---

[i]Standard bull-semen type ampules, Wheaton Cryule No. 12483 (unscored) and No. 12523 (prescored); Special wide-mouthed ampules, Wheaton Gold Band Cell and Tissue Cryule No. 12742; Celstir Unit No. 356676-79 Wheaton Scientific Co., Millville, New Jersey 08322.

[ii]Markem Corporation, 150 Congress St., Keene, New Hampshire 03431.

[iii]Nasco (No. A545) Fort Atkinson, Wisconsin 53538.

Globe semi-automatic ampule sealer[iv]. This unit is designed to seal ampules using the pull-seal method. The pull-seal method is preferred over the tip-seal method because it minimizes the occurrence of pin-hole leaks in the tip of the ampule. After sealing, the ampules are tested for leaks in a dye solution as mentioned above. Such safety procedures obviate the potential hazard of improperly sealed ampules exploding when they are rapidly removed from the liquid nitrogen and are placed at room temperature. This hazard can also be circumvented by storing the ampules in the vapor phase of a liquid nitrogen refrigerator.

For freezing the cells we routinely use a Linde BF3-2 freezer[v] (which operates on a differential thermocouple principle). We use this unit because the freezing rate can be easily varied from 1/2 to approximately 20 degrees per minute. For critical experiments and repository production procedures controlled freezing rates are quite necessary to provide permanent records of freezing curves and to insure reproducibility. In other instances, however, simpler procedures are just as suitable. For instance, many cells can be successfully frozen by simply placing the ampules in a dry ice chest or mechanical refrigerator at -65 C for several hours, or even overnight. This provides an adequate rate of freezing then, after the cells are frozen, they may be placed in a liquid nitrogen refrigerator for long-term storage.

For recovery of the cells from the frozen state the following general procedure is followed. The cells are thawed by removing the ampule rapidly from the liquid nitrogen refrigerator and plunging it immediately into a water bath at 37 -40 C. A higher temperature, 41 - 45 C, is required for plant rusts[8,9]. The thawing of the cells should be accomplished as rapidly as possible (with moderately vigorous agitation, the ice usually melts within 40-60 seconds). Immediately after thawing, the ampules are removed from the water bath and immersed in 70% ethanol at room temperature.

With animal cells it is particularly important to avoid excessive alkalinity of the culture medium during recovery of the cells from the frozen state. The pH of the medium should be adjusted to approximately 7.2 *prior* to the addition of the ampule contents. In recovering animal cells we find it best (although not absolutely necessary) to introduce the ampule contents diluted with approximately 10 ml of fresh culture medium into a single flask. With cell lines that have a high minimum inoculum size this helps to insure the establishment of a vigorously growing culture. The 10-fold dilution of the ampule contents lowers the

---

[iv] K-6 ampoule sealer Bench Model 161 Kahlenberg-Globe Equipment Co., P. O. Box 2803, Sarasota, Florida 33578.

[v] Union Carbide Corporation, Linde Division, P. O. Box 766, New York 10019.

concentration of the cryoprotective agent to a level that does not necessitate its complete removal by centrifugation. In order to expedite the removal of the freezing additive the culture medium should be changed the day after the cells are thawed. With the bacterial and fungi the material in the ampule is simply streaked out on agar slants or plates or inoculated into broth cultures.

Employing the procedures described above the ATCC prepares 10 to 100 ampules of each strain to serve as frozen reference seed stocks for long-term preservation. The number prepared depends upon the requirements of the specific collection. Additional ampules, that serve as distribution stocks, are then prepared from the reference seed stocks as needed. With particular reference to the animal cells all quality control tests and characterizations are performed on aliquots or cultures derived from the reference seed stock at a *specific* passage level. Thus the availability of such reference seed stocks from a public repository provides researchers with a *common* source of characterized cells today, tomorrow, or even many years from now. This reference seed stock concept increases the possibility that meaningful comparative studies can be made between different laboratories throughout the world and also between different generations of investigators. The ultimate advantages of working with characterized cells "captured" at a specific stage in their life history cannot be overemphasized. The banking of cells in liquid nitrogen also makes possible the collection and long-term storage of rare cells or cells that for other reasons are difficult to acquire; for instance, cells from certain wild animals[5] or cells from patients with rare diseases[17].

REFERENCES

1. Stulberg, C.S., W.D. Peterson, Jr. and L. Verman, 1962. Quantitative and qualitative preservation of cell-strain characteristics. Nat. Cancer Inst. Monograph *7:*17-32.

2. Evans, V.J., H. Montes De Oca, J.C. Bryant, E.L. Schilling and J.E. Shannon, 1962. Recovery from liquid nitrogen temperature of established cell lines frozen in chemically defined medium. J. Nat. Cancer Inst. *29:*749-757.

3. Greene, A.E., B. Athreya, H.B. Lehr and L.L. Coriell, 1967. Viability of cell cultures following extended preservation in liquid nitrogen. Proc.Soc.Exp.Biol. and Med. *124:*1302-1307

4. Stulberg, C.L., L.L. Coriell, A.J. Kniazeff and J.E. Shannon, 1970. The Animal Cell Culture Collection. In Vitro *5:*1-16

5. American Type Culture Collection, Registry of Animal Cell Lines, Second Edition, 1972. (J.E. Shannon and M.L. Macy, Editors).

6. Hwang, S.W., 1966. Long-term preservation of fungus cultures with liquid nitrogen refrigeration. Appl. Microbiol. *14:*784-788.

7. Hwang, S.W., 1968. Investigation of ultra-low temperature for fungal cultures. I. An evaluation of liquid nitrogen storage for preservation of selected fungal cultures. Mycologia *60*(3):613-621.

8. Loegering, W.Q., H.H. McKinney, D.L. Harmon and W.A. Clark, 1961. A long term experiment for preservation of urediospores of *Puccinia graminis tritici* in liquid nitrogen. Plant Dis. Rep. *45:*384-385.

9. Loegering, W.Q. and D.L. Harmon, 1962. Effect of thawing temperature on urediospores of *Puccinia graminis* var. *tritici*. Phytopathology *46:*299-302.

10. Shannon, J.E., E.B. Franck, S.W. Hwang and M. Macy, 1970. Effects of storage temperature on the long-term viability of selected strains of microorganisms and mammalian cells. Presented at the Xth International Congress of Microbiology, Mexico City.

11. Clark, W.A., 1970. The American Type Culture Collection: Experience in Freezing and Freeze-Drying Microorganisms, Viruses, and Cell Lines. *In* Proceedings of the First International Conference on Culture Collections, (Tokio Nei, ed.). University of Tokyo Press, Tokyo, Japan.

12. Clark, W.A., 1962. Attempts to freeze some bacteriophages to ultra low temperatures. Appl. Microbiol. *10:*463-465.

13. Clark, W.A., and D. Geary, 1973. Preservation of bacteriophages by freezing and freeze-drying. Cryobiol. *10:*351-360.

14. Norman, M.C., E.B. Franck and R.V. Choate, 1970. Preservation on *Mycoplasma* strains by freezing in liquid nitrogen and by lyophilization with sucrose. Applied Microbiol. *20*(1):69-71.

15. Hwang, S.W., E.E. Davis and M.T. Alexander, 1964. Freezing and viability of *Tetrahymena pyriformis* in dimethylsulfoxide. Science *144:*64-65.

16. Hwang, S.W. and W. Horneland, 1965. Survival of algal cultures after freezing by controlled and uncontrolled cooling. Cryobiology *1:*305-311.

17. American Type Culture Collection, List of Human Skin Fibroblasts (Including Genetic Disorders, Other Disease States, and Normal Controls). Third Edition, 1974. (J.E. Shannon, Editor).

18. American Type Culture Collection, Catalogue of Strains, Eleventh Edition, 1974. M.T. Alexander, W. A. Clark, E. Davis, H. Hatt, S.C. Jong, E.F. Lessel and R. Zieg, Editors).

19. American Type Culture Collection, Catalogue of Viruses, Rickettsiae, Chlamydiae, Fourth Edition, 1971. (T.O. Berge and D.A. Stevens, Editors).

20. Shannon, J.E. and M.L. Macy, 1973. Freezing, storage and recovery of cell stocks. In *Tissue Culture: Methods and Applications*, Edited by P.F. Kruse, Jr. and M.K. Patterson, Jr., pp. 712-718, Academic Press, New York.

SOME CONSIDERATIONS IN THE CRYOGENIC PRESERVATION OF CULTURES
by
William H. Trejo
Department of Microbiology
The Squibb Institute for Medical Research

In considering the choice of any system for processing cultures, thought should be given to the function that the culture collection is to serve. It matters little whether we are talking about national repositories, specialist collections or industrial collections because, essentially, the prime functions of all collections are:

1. preservation of cultures without loss of any morphological, physiological or biochemical properties (in short, retention of their genetic integrity); and

2. distribution of cultures.

The extent of the distribution service directly influences the method of preservation and storage. Lyophilization, for example, despite the associated high frequency of killing of cells, is still used by many major collections because lyophilized cultures are better suited for distribution than are ampoules stored in liquid nitrogen (LN_2). Cultures in ampoules stored in LN_2 must generally be transferred to solid media before shipment; although ampoules in LN_2 may be shipped in special containers, this procedure is costly and risky.

Our collection at Squibb is representative of most industrial collections. We maintain microorganisms for assay purposes, research in taxonomy, antibiotic biosynthesis, studies on mechanism of action of antimicrobial agents, biochemical transformations, and studies in chemotherapy. Distribution of cultures is primarily within Squibb or to our overseas affiliates, but we exchange cultures with other collections, institutions, or individuals on a courtesy basis. As a matter of policy of supporting the American Type Culture Collection (ATCC), we do not distribute cultures that are available through that collection. It would erode the financial position of the ATCC for other collections to distribute free cultures for which they must charge a fee. The sale of cultures and culture-related services is important to the scientific community and provides a source of income to the collection.

We began using preservation with LN_2 in 1963 and no longer preserve cultures by lyophilization. We maintain between 2,000 and 3,000 cultures, about 50% of the number we had two years ago. Because preservation in LN_2 is effective and convenient, investigators often tend to deposit in the collection, for future use, cultures of only minimal interest. To avoid this, we review the status of all cultures periodically to determine whether continued storage is warranted.

Our collection may be subdivided as follows:

1. actinomycetes, bacteria, fungi, and yeasts used largely for screening;

2. reference cultures, mainly type or neotype cultures, used for taxonomic studies;

3. organisms used for the assay of antibiotics, vitamins, etc.;

4. cultures cited in company patents or publications; and

5. a large number of fresh clinical isolates, mostly bacteria, used for documenting the activities of antibiotics *in vitro*. Normally, we maintain a limited number of ampoules of each culture for an unlimited term of storage. Storage of this group poses a set of problems different from those in groups 1-4, because it is necessary to prepare a sufficient number of vials of each culture to maintain working stocks for the testing group. Since these cultures are rarely kept longer than two years, long-term survival is of little concern.

PROCEDURES

In preparing cultures for preservation, it is essential that the cells be in a proper state for optimal survival. Nonsporeforming bacterial cultures should be in the active logarithmic phase. Spore-formers should be harvested at the peak of sporulation, as determined by microscopic examination. Actinomycete and fungal cultures should be well sporulated and given sufficient time for spore maturation. Spores develop in these organisms within 7-10 days, but full maturation is not reached till 12 or more days. In the case of nonsporulating fungi, a vigorous vegetative mycelium scraped off a slant or grown in shaken culture may be used. Spores or cells are washed off the surface of slant cultures. Although no cell counts are made routinely, we are dealing with a massive inoculum, perhaps 10^8 or 10^9 colony-forming units per milliliter. These cells are suspended in a suitable cryoprotective agent as a menstruum, and the ampoules are sealed in a gas-oxygen flame.

Cryoprotective agents are of two types:

1. penetrating agents - glycerol, dimethylsulfoxide; and

2. nonpenetrating agents - sucrose, lactose, glucose, mannitol, sorbitol, dextran, polyvinyl-pyrolidone and polyglycol.

We use 7% glycerol in water exclusively, and have had no experience with nonpenetrating agents.

CONTROLLED COOLING RATES

Much has been written about the effect on survival exerted by the cooling rate of cells during freezing. In our work, we routinely use a cooling rate of 1 degree per min which has worked well for us even with such fastidious organisms as *Neisseria*, *Hemophilus*, and *Brucella*. Other cooling rates are used for the preservation of tissue cell lines, protozoa, and some algae.

We have used both the simple plug freezer Linde BF-3 and the much more sophisticated Linde BF-4 controlled-rate freezer to obtain controlled cooling rates.

The plug freezer, which holds nine ampoules, controls the cooling rate by regulating the depth of penetration of the plug into the neck of the Linde LR-35 refrigerator by positioning the "O" ring at different stops. The deeper the penetration, the more rapid is the cooling rate. By carefully maintaining a uniform level of nitrogen in the LR-35 during each run, one can come surprisingly close to a rate of 1 deg/min by following the simple instructions provided by Linde. The chief disadvantage of this method is the limited number of ampoules that can be processed at one time. However, when a small number of ampoules are to be frozen this system is simple and economical. We operated for quite a few years with five LR-35s and plug freezers. The cost of a plug freezer is about $36.00.

The Linde BF-4 controlled-rate freezer was acquired at a time when we were involved in tissue culture work. This elegantly engineering device is extremely precise and has a 64-ampoule capacity. The ampoules are placed in an insulated cooling box into which LN_2 is fed in short bursts through a solenoid valve. An electronic controller programs the timing of the nitrogen feed to maintain any pre-set cooling rate. The temperature of a sample ampoule (uninoculated menstruum) is monitored throughout the run by means of a thermistor and a recording thermograph. When the run has been completed, the ampoules are quickly transferred to aluminum canes, which are then stored under LN_2. In our experience, however, the long-term viability of cultures cooled with the plug freezer is as good as that of cultures cooled more precisely with the BF-4. Unless your laboratory is involved in extensive preservation of tissue cultures, protozoa, or intact organs, we cannot see the justification for a BF-4 freezer. The cost of a BF-4, complete with recorder, is about $1,800.00.

Some laboratories report successful recovery of cultures immersed directly in LN_2. Other laboratories are cooling ampoules in styrofoam blocks placed in -70 C freezers, and have reported a 2.5/min cooling rate in the critical temperature range.

LN_2 REFRIGERATORS AND USAGE

Most LN_2-refrigerators can be operated either as liquid-phase or vapor-phase refrigerators. The chief advantage of vapor-phase storage is safety. Storage in liquid-phase refrigerators requires proper sealing of the ampoules that are to be immersed in LN_2 (-196 C), for the slightest leak permits entry of nitrogen into the ampoule. During rapid thawing at 37 C, improperly sealed ampoules invariably explode and create a serious hazard. No such problem exists in vapor phase (-100 to -150 C) storage, for screwtop ampoules may be used.

One of the most important considerations in the management of culture collections is the organization of the stored cultures. Not only does the organization affect the location and retrieval of cultures, it also affects the consumption of liquid nitrogen. Two methods of storage commonly used are:

1. vertical storage, in which 5-6 ampoules can be clipped to an aluminum cane, thus in our system, 16 canes are clustered and stored vertically in baskets holding 12 clusters per basket; and

2. drawer storage, in which the ampoules are stored in stacked drawers or boxes.

In our experience, vertical storage is far more efficient in the use of space and ease of retrieval. The reference position of each culture in the freezer should appear on each culture record card. The ampoules and canes should be clearly and permanently labelled. It is important to keep a running inventory of stocks, and stocks should be replenished regularly to minimize the storing of empty or near-empty canes. It is surprising how fast empty canes can accumulate in a busy collection.

Since cultures are recorded in the order of acquisition, it would seem to be logical to store them in the same order. Numerical sequence, however, is rapidly lost in any large collection, particularly as cultures are discarded and their assigned numbers are not used again. We find that it is better to adjust both the number of ampoules of a culture and their location in the freezer according to the frequency of demand for the culture. Thus, cultures not required frequently are stored in the less accessible sections of the refrigerator. We are also storing together cultures that are usually requested together. These practices reduce the frequency and duration of opening of the refrigerator.

In the system that uses preservation with LN_2 the greatest expenditures will be for the refrigerator and the cost of the nitrogen itself. Therefore, some thought should be given to overall storage capacity, static evaporation rates, and static holding times. These parameters have some bearing on overall nitrogen consumption, but are not the sole determinants.

Static evaporation rates and static holding times are indicators of how well the equipment is insulated. They determine the maximum time between fillings in an *unopened* refrigerator, but do not take into account the normal operational use. The total consumption of nitrogen is determined primarily by the design of the equipment used.

The chief factors contributing to loss of LN_2 are:

1. size of the opening of the storage unit;

2. frequency of opening;

3. duration of the open period; and

4. ambient temperature of the room.

The LR-310 refrigerator requires about 56 liters to maintain a 3-4 inch level of LN_2 in the bottom of the tub for vapor-phase storage. An automatic filling system controlled by a solenoid valve delivers nitrogen from the reservoir. The opening and closing of the solenoid are controlled by thermistors, and the whole assembly is attached to the nitrogen entry line. Additional thermistors actuate an alarm system if the solenoid fails to open or close.

The LR-310 has a static holding time of eight days, but our two units have failed to hold during a long weekend. To correct this, we readjusted the position of the sensing thermistors to maintain a 10-inch level of LN_2 in the tub. This modification, which required about 200 liters of LN_2, effectively made the refrigerator a combination liquid-vapor-phase storage unit. In this way, we avoided the risk of ampoules thawing and were no longer subjected to anguished telephone calls at 4 A.M. from plant security guards alerted by the alarm system.

We have encountered problems with the solenoid sticking either in the open or closed position or short cycling on and off in a manner seemingly independent of the thermistor controls. Replacement of the solenoid helped for a time, but the problem returned. We found that considerable turbulence is set up at the nitrogen inlet because of the supporting tray on the bottom of the tub. This turbulence seemed to travel up the sensing probe, tripping the solenoid closed even though the preset level had not been reached. Much of the wear of the solenoid from short-cycling could probably be ascribed to this, as could come of our losses of nitrogen.

As mentioned previously, the LR-310 has about twice the storage capacity of the LR-250 because of a deeper tub that permits two-tiered storage. The chief disadvantage of two-tiered storage is that it takes extra time to get down to the bottom tier, thus increasing the duration of open time. The 29-inch diameter refrigerator opening also increases nitrogen loss. The LR-250, in contrast, provides single-tiered storage and the lid is divided, so that only half of it need be opened at any time. This arrangement effectively reduces the size of the opening of the storage unit and the duration of open time, thereby conserving nitrogen.

Under our operating conditions, we found that the LR-310 refrigerator consumed more nitrogen than did the LR-250, taking into account the relative storage capacities of the two units. The consumption and costs are summarized in the following table:

<u>Comparative Consumption of LN_2 by LR-250 and LR-310 Refrigerators</u>

	LR-250	LR-310
Manufacturer's estimated consumption (liters/month). (See discussion of static rates in text.)	174	270
Actual consumption	320 x 2* (640)	960
Cost/month**	$224	$336

(*) Correction for relative storage capacity of the two units.

(**) Based on $0.35/liter delivered in LS-160 transport cylinders.

The manufacturer's estimated consumption of LN_2 for both units seems too conservative. Admittedly, actual consumption will vary from one laboratory to another, but few operate as dead storage. Any estimate of consumption should allow for a realistic amount of operational open time in the refrigerator.

Any installation that is using about 1,000 liters (*ca* 25,000 ft^3) of nitrogen or more per month should give serious consideration to bulk storage of LN_2. The saving of buying in bulk is substantial, because of reduced transportation and handling costs. Bulk storage facilities may be purchased outright or rented. A minimum monthly volume of 20,000-25,000 ft^3 (1 liter = 28 ft^3) is required by the distributor. The cost of nitrogen may range from 3-12¢/liter. Bulk storage should be located close to the area that is to be serviced, otherwise in plant transportation costs or capital outlay for expensive vacuum-insulated feed lines will cancel any savings in cost of nitrogen.

Should an investment in bulk-storage facilities be ruled out because of the high cost, then alternative methods have to be considered. For us, the alternative has been mechanical refrigeration and, at present, we have three Revco (ULT-695) units in operation.

<u>MECHANICAL REFRIGERATION</u>

In terms of storage capacity, each Revco refrigerator holds approximately the same number of cultures as does the Linde LN_2 refrigerator, LR-250. Although the operating temperature is only -95 C, we find this to be adequate for a great many cultures, particularly our collection of clinical isolates.

In these times of energy crisis, the danger of blackouts or brownouts with normal electrical service is a serious concern. Fortunately, these units are wired into an emergency reserve power (ERP) source that comes on within seconds of failure of the normal service. Mechanical refrigeration should not be considered unless there is an emergency power service available. Damage to compressors can result from the combination of low voltage or complete loss of power and a sudden start-up when power is restored.

In two years of operation, we have had three compressor failures, unrelated to power failure, and two alarm circuits burned out. For such an eventuality, it is advisable to have either an empty standby mechanical freezer running or liquid nitrogen available to put into such a freezer until the compressor has been repaired. It has been our experience that, when a compressor fails, the unopened refrigerator will not maintain its temperature very long. The stored material may remain frozen, but the temperature will rise to levels in which biochemical changes and cell damage occur (-60 to 0 C).

We consider storage in LN_2 to be the method of choice for the long-term preservation of microorganisms. The greater temperature spread (-196 to -60 C) provides a wider margin of safety than that obtained with mechanical refrigeration (-95 C to -60 C).

We recognize, however, that cost considerations in some laboratories may make mechanical refrigeration a viable alternative to preservation with LN_2. Selection of equipment should be tailored to the specific needs of the laboratory. It may be that a combination of storage methods is most suitable, depending on the types of organisms stored and the function of the collection. Proper organization of the storage unit is essential to minimize operational costs.

PRESERVATION AND SHIPPING OF INDUSTRIAL STARTER CULTURES

Robert L. Sellars
Chr. Hansen's Laboratory, Inc.

This conference was formed with the basic intent to disseminate pertinent information concerning the use of cryogenics -- primarily liquid nitrogen -- as an effective and practical means of preserving biological materials. The following, therefore, is a discussion of some of the practical problems which were involved during the research and development of a particular type of commercial product, commercial lactic acid starter-cultures. Perhaps by relating some of our particular experiences, positive as well as negative, those of you who have an interest in cryogenic preservation of biologicals but have been somewhat reluctant to pursue it for various reasons, may receive some benefits from our results found to be applicable to lactic starter cultures, their preservation, mode of storage and method of distribution.

A brief description of the type of biological material which has been effectively preserved and commercially available for the past eight years is in order. To receive a better understanding of this material and how it is commercially applied, a brief historical background as to the nature and use of lactic acid-producing starter bacteria as they relate to the various sectors of the food industry is presented.

Lactic acid-producing organisms normally found in industrial starter-cultures have been available commercially and used throughout the dairy industry for the manufacture of various types of cheeses since the early nineteen hundreds. While some companies in the processed meat sector have been using starters since the middle to late '50's, the use of industrial cultures for controlled fermentation of processed meats has not been universally accepted. However, the development of concentrated, cryogenically frozen, and preserved material has increased their use. A significant portion of processed meats of summer, dry and semi-dry, type sausages, bolognas, cervelats, thuringer, and the like is now manufactured under controlled conditions with added starter-cultures. In 1973 pickles and other pickled products (relish, peppers, etc.) were commercially produced under controlled procedures by the addition of highly concentrated cryogenically frozen and preserved starter-cultures. More recently, cryogenically frozen and preserved starter-cultures became available to the baking industry for the manufacture of sourdough bread. Their acceptance and use to date is steadily increasing. New applications are being developed weekly for baked goods other than bread. Their potential use in many of our bakery products appears unlimited.

The use of starter-cultures *may* reduce the necessity of adding preservatives under given conditions. Improved cryogenic technology has made possible the use of starter-cultures in foods and fiber not previously possible.

The principal mesophilic strains which compose lactic starter-cultures for the dairy industry belong to the genuses of Streptococci (namely, lactis, cremoris, diacetilactis, and the thermophilic species of *S. thermophilus*). Selected strains of Lactobacilli (namely, bulgaricus, helveticus, and casei) are also included. The Lactobacilli strains plus the *S. thermophilus* strains are used for the manufacture of yogurt and Italian cheese varieties, whereas the Streptococci plus selected strains of *Leuconostoc citrovorum* are used for the manufacture of American type cheeses, cultured buttermilk, cultured sour cream, cottage, cream and bakers cheese. While strains of Leuconostoc and *Propionibacterium shermanii*, used in the manufacture of Swiss cheese, do not produce appreciable amounts of acids, they are included in the group of commercial starter-cultures which are effectively preserved cryogenically.

Commercial "starter-cultures" have been available from Chr. Hansen's Laboratory, Inc. since 1893. The first cultures were mixtures of various types of unidentified, but reasonably pure strains of organisms which were, years later, to be classified as *Streptococcus lactis*, *Streptococcus cremoris*, and species of Leuconostoc.

The first commercial starters were available only as a "mother" culture which was simply a milk-fermented, acid-coagulated medium. They were sold in quart glass bottles. At that time distribution was not a particular problem since 80% of the business was confined principally to a small geographical area, primarily New York and Pennsylvania. As the art of cheese-making spread across the country and the demand for these starters increased, it became necessary to add calcium carbonate which helped to neutralize the acids developed during fermentation. This rather simple, but effective addition increased the "shelf-life" of these starters by several days. As the dairy industry continued its unparalleled growth during the late 1920's and 1930's, and as our U. S. Postal System began to offer more efficient and greater coverage, the cultures were produced by a unique process of drying that eventually led to the development of a freeze-drying system which even today maintains excellent recovery of the initial viable organisms with superior activity (Note: Excellent preservation is maintained in this lyophilized form for more than twenty years when stored at -20 C). The dry, and later the freeze-dried preparations were most effective as a means of preserving these organisms in a way in which they were used all over the world. The freeze-dried form is still being used effectively at the present time in many of the South American, African, and some European countries.

In the early to middle 1950's serious problems of bacteriophage infections of these starter-cultures developed which nearly crippled the dairy industry. Infections were occurring, primarily because of the increased volume of manufacturing; because of the particular procedures used by the cheesemaker in the preparation of his "bulk-culture." i.e., that culture material which is added to the cheese milk usually at the rate of 1% w/w; because of the changes in the microflora of the raw milk; and because of the changes in handling and processing of the milk, cheese whey, and the cheese itself.

The normal procedure employed for eventually preparing the "bulk starter--culture" in the dairy industry was to prepare a mother culture, usually in quantities of one pint to one quart. Then, an intermediate batch which ranged in capacities of 10 to 50 gallons was prepared which was then used to inoculate 100 to 500 gallons of bulk starter. It was during these steps that the bacteriophage or viral infection was occuring. Several different approaches were developed and used to reduce this serious problem. None of these were significantly effective until the development of the phosphate phage-protective media and its application as outlined in the steps above. The use of phosphate phage-protective media has effectively reduced the incidence of phage infections. However, the initial application of this media still involved basically the preparation of mother and intermediate cultures. By using cryogenic techniques, it was recognized that if a suitable container could be found which would meet the necessary specifications, then the bulk culture could be prepared by direct inoculation. Today, approximately 90% of all bulk starter-culture is prepared by the "direct bulk set cultures". However, more recently, the development of a "Direct Vat Set" culture (a super concentrated culture) is used to inoculate the ultimate consumer product such as the cheese milk itself. This step eliminates the need for starter-culture preparations. (Note: In meat and bakery products the starters are added directly to the ultimate consumer product, whereas, in vegetable controlled fermentation the starter-cultures are used to inoculate bulk quantities of 200 to 12,000 gallons of brined material.)

During the early stages of development, extreme difficulties were experienced in finding a suitable and satisfactory container which met the specifications set forth as dictated by the availability of cryogenic hardware used for freezing, storage, and distribution and by the individual characteristics of lactic acid starter-cultures.

The specifications which any container for this type of commercial product had to meet were as follows: It had to: 1) be non toxic to the organisms; 2) be approved by FDA; 3) be resistant to acid; 4) be completely sealable; 5) be sterilizable by heat; 6) be adaptable to the cryogenic refrigerators which were commercially available; 7) be non hazardous to the handler or user; 8) be easy to handle during production and when used by the user; 9) be capable of holding a sufficient quantity of a concentrated material; and 10) retain its integrity during storage at the cryogenic temperatures of liquid nitrogen and/or its vapor. While these are rather stringent requirements, they were absolutely necessary. These specifications are still valid at the present time.

The primary problems in developing a direct inoculum were: 1) to effectively and efficiently concentrate a highly viable and active culture material *at an economical cost;* and 2) to find or develop a suitable package or container to hold this product. Both were somewhat dependent upon the other. For example, a highly concentrated culture was of no value if the container or package was a poor conductor of heat. The culture menstruum had to be frozen as rapidly as possible. The rate of freezing was found to be influenced to a great degree by the nature of the material frozen. Earlier attempts to employ cryoprotective agents were unacceptable. Certain cryoprotective agents can now be used very effectively. Some of the various types of containers and packages which were representative of the basic container types are presented below.

The sealable 1.0 and 1.2 ml glass ampoule was used in our earlier studies to measure the rates of freezing and freeze-thawing in liquid and nitrogen vapor. This type of container was also used to measure the effects of freeze-thaw on the *viability* of the different concentrated materials. During these earlier studies not all lactic Streptococci and Lactobacilli gave the same percentage response in viability and activity (the rate of acids produced by individual cells) to the freeze-thaw procedures. This fact compounded the problem of reaching the ultimate goal. Glass ampoules were found to be satisfactory for storage of stock culture, for storage of small quantities of material and for studying the rates of freeze-thawing. Since this type of container presents some safety hazards, they can and should be handled with gloves and safety goggles, particularly when the frozen cultures have been stored under liquid. However, they were not commercially acceptable for large volumes of product.

The screw-capped ampoule was found to be quite satisfactory for storing quantities of material up to ten milliliters and for material which did not have to be frozen in liquid nitrogen since at that time (Wheaton Glass Co.) they were not leakproof. More recently, a screw capped ampoule which is leakproof is available. The rate of freeze-thaw was found to be comparable to the sealable ampoule. Also there were no safety hazards involved. They make excellent containers for preserving enzyme preparations for seed culture inocula or for samples which are extracted during an enzyme reaction sequence.

The aluminum R-2 pouch (Continental Can Co.) will retain its integrity up to several months in the liquid and somewhat longer in the vapor phase. The time varies somewhat depending upon the method of freezing, storing, and thawing. The freeze-thaw rates are, however, a little slower than in the glass ampoule. This type of package has one advantage in that it can be printed for easy identification. Another distinct advantage is that it can be fabricated to almost any size, which permits greater flexibility in the quantity of material one can store.

The clear plastic type bag -- called Bioflex (Union Carbide) -- has several advantages over the R-2 pouch and other flexible materials. This material will retain its integrity for at least two years in both liquid and vapor which for all practical purposes is commercially feasible. The rates of freeze-thaw are comparable to glass and slightly better than the R-2 pouch but not as good as the aluminum can. The Bioflex material has been used for the storage of blood. It is easily sealed; is less likely to develop leaks; and can be fitted with ports, etc. or made into different geometrical shapes -- round, square, and rectangular. The main disadvantage of this package at the present time is the cost which will most likely be improved as more uses are found for it. Of all packages and containers of this type which were tested, this Bioflex material has the most desirable cryogenic characteristics. Presently, this material is not being used for distributing commercial starters; however, in the future as the quantity requirements increase for use in the food industry, this type of container could become more attractive.

The drawn aluminum cans (Continental Can Co. and Central States Can Co.) have mainly two disadvantages; 1) they will sometimes leak because of improper seals; and 2) the largest size presently available is the 15 oz. size. Four sizes are available -- 1.5, 3, 6, and 15 ounce. Only the three larger sizes are currently being used for the distribution for our type of industrial starter-cultures. Cryogenic refrigerators are currently available to handle all three sizes easily and conveniently for storage and distribution. (Linde Division, Union Carbide Corp., and Minnesota Valley Engineering).

These aluminum cans have been used now for more than eight years. They have other advantages and desirable characteristics. They are light weight, can be printed, are fairly rugged, and are not susceptible to puncture. They can be easily opened by the customer with the peel-type lids and are economically attractive. The units meet all specifications as outlined previously, whereas many of the other containers examined did not qualify.

The advantages of cryogenic preservation of lactic starter-cultures and the technical differences between liquid cultures, freeze-dried, and nitrogen frozen cultures have been published previously (Microbial Technology, edited by H. Peppler). Briefly, though, cryogenic preservation offers the best method to date of preserving starters at their maximum viability and activity by flash freezing in nitrogen which subjects the organisms to what has been referred to as "suspended animation." Storage up to ten years without any significant change is possible. Biological material of this type can be preserved effectively for much longer periods. Many cultures after ten years of storage were found to be just as good and in some cases slightly better than they were just prior to or immediately after flash freezing. For commercial purposes two years is more than sufficient.

The development of direct inoculum starter-cultures has eliminated the extra handling procedures which were previously necessary in the preparation of starter-cultures normally used in the ultimate consumer product. They have effectively reduced the incidence of phage infections in the dairy industry and thereby saved millions of dollars in potential losses. They have not, however, completely eliminated phage infections. The concentrated, highly active cultures have also given more consistent and uniform performance, thereby more uniform products; have offered better opportunities for quality control procedures, and have been more economical for the customer.

The cryogenic frozen starters for meat, vegetable, and bakery products have also eliminated extra handling procedures. However, their effective use also requires more technologically controlled fermentation procedures under more sanitary conditions. Less product losses occur while maintaining greater uniformity not possible by "natural fermentations" commonly used in previous years.

This development would not have been possible without the cooperation of the suppliers of nitrogen cryogenic equipment, packages, and containers. Regardless of the type of package or containers selected or studied the methods of storage and distribution had to be kept constantly in mind. At times a suitable container met all criteria and specifications but did not meet the distribution requirements. Eventually, all of this was solved and has opened the door to future potentials which would not have been possible otherwise.

This report is only an overview of some of the problems encountered during this commercial development. Perhaps some of it has been helpful and of some interest.

NITROGEN PRESERVATION OF STOCK CULTURES OF UNICELLULAR AND FILAMENTOUS MICROORGANISMS

Alma Dietz

Infectious Diseases Research
The Upjohn Company

Microbial stock cultures must be maintained so that the desired metabolic and morphological properties exhibited prior to storage are retained by subcultures of the stored organisms. In 1961, in an article entitled "Cryobiology," Dr. Joseph F. Saunders[1] predicted long-term preservation and storage of microorganisms by cryogenic methods and the concomitant advantages of stabilization of the specimens and elimination of laborious maintenance methods. Dr. Saunders' predictions for cryogenic storage of microorganisms have been realized in The Upjohn Company Laboratories. Stock cultures of unicellular and filamentous organisms are maintained in the gas phase of liquid nitrogen refrigerated storage tanks. In this paper simple procedures are given for reliable storage, retrieval, and reconstitution of such microorganisms.

CULTURE GROWTH AND STORAGE METHODOLOGY FOR NITROGEN PRESERVATION OF MICROORGANISMS

Unicellular Organisms

Bacteria. These organisms are grown in or on the medium of choice in culture tubes for 18-24 hours at the optimum temperature. Aliquots of broth cultures are distributed into storage vials. Sterile distilled water is added to slant or stab cultures and suspensions of the growth are made in distilled water. Aliquots of the suspensions are distributed in storage vials. Vials and storage conditions employed in The Upjohn Stock Culture Collection are shown in Figures 1-5.

Figure 1 illustrates the preparation of vials for storing bacteria. One milliliter of inoculum is dispensed in a 1/2 dram screw cap vial (Kimble* No. 60910, 12 x 35 mm). Vial caps must be foil-lined.

---

[1]Saunders, J. R. (1961) Cryobiology. Naval Research Reviews.

*Reference to equipment used and cited by manufacturer's number is not an endorsement of that manufacturer. Once a satisfactory procedure has been established it is not an expedient use of time to explore all products of potential use.

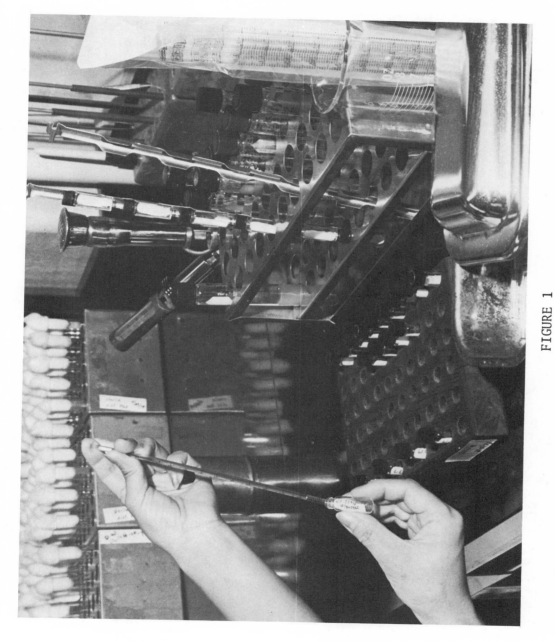

FIGURE 1

Preparation of Vials for Bacterial Storage

Paper-lined caps are cumbersome; the paper adheres to the vial top after autoclaving and must be removed before a vial can be inoculated.

Figure 2 shows a Union Carbide LR-640C liquid nitrogen refrigerator. This unit is adjusted for refrigeration in the gas phase of liquid nitrogen. A liquid level is maintained beneath the storage area by automatic filling from an LS-160 supply tank (Figure 3). This tank has a capacity of 160 liters of liquid nitrogen with a gauge pressure of not more than 22 pounds (22 psig). The supply tank is fitted with a Kamlok-Quick Disconnect Coupler (633-D) obtained from OPW Division of Dover Corporation, Cincinnati, Ohio. Transfer of nitrogen is effected with a standard Linde metal bronze hose (1191-9C95). Figure 4 depicts the ease with which the flexible hose is disconnected from the supply tank.

Figure 5 shows part of the interior of the LR-640C. The unit has a two-tier lazy-Susan arrangement of pie-shaped sections. Pie-shaped storage inserts for each section were fabricated in The Upjohn Company Sheet Metal Shop. The units are three-tiered. The top and middle tiers have holes spaced to accommodate 335 canes. (The canes are available from Shur-Bend Mfg. Co., Inc., Minneapolis, Minnesota.) Each cane holds 6 1/2 dram screw cap vials. One cane is used for each culture stored. A replica of the insert is drawn on heavy cardboard. This is used to record the location of the cultures stored and for retrieval.

Algae, Phages, Protozoa, Yeasts. Procedures for growing these organisms are the same as for filamentous organisms described below.

Filamentous Organisms

Actinomycetes and Fungi. These organisms are grown on the agar medium of choice in a petri plate for the desired incubation period. Agar plugs of the growth are then stored in straws in screw cap vials. (Cultures with extreme aerial growth are more easily plugged as soon as compact surface growth is noted.) Procedures employed in The Upjohn Stock Culture Collection are shown in Figures 6-9.

Figure 6 illustrates the preparation of vials for agar plugging. The procedure developed at The Upjohn Company is as follows: ELBO flexible drinking straws (paper), Item No. 308 manufactured by National Soda Straw Company, Chicago, Illinois, are used. Five pieces are cut from each straw. (The flexible section is discarded.) Each piece has a flat end for plugging and a slanted end for insertion of a tapered holder for leverage in plugging. Five straw pieces are used per screw cap vial (Kimbal No. 60957-88, Size 27 1/4MM x 2").

Figure 7 shows the Union Carbide LR-1000 liquid nitrogen refrigerator. This unit is also adjusted for refrigeration in the gas phase of liquid nitrogen. The temperature indicator shown on the tank has a range of -200 to +100C. It is Thermocouple Type T (cooper-constantan) obtained from Cryo-Med, Detroit, Michigan.

FIGURE 2

Union Carbide LR-640C Liquid Nitrogen Refrigerator

FIGURE 3
LS-160 Liquid Nitrogen Supply Tank

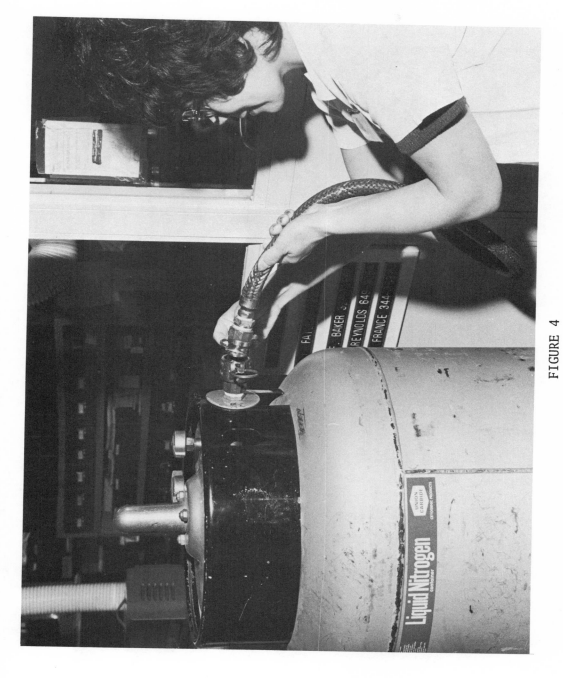

FIGURE 4

Kamlok Quick Disconnect Coupler and Linde Metal Bronze Transfer Hose

FIGURE 5

Part of Interior of LR-640C Liquid Nitrogen Refrigerator

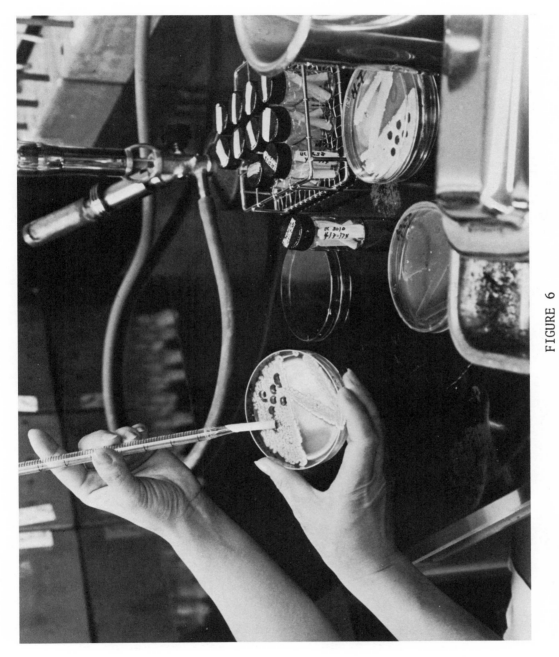

FIGURE 6

Preparation of Vials for Agar Plugging

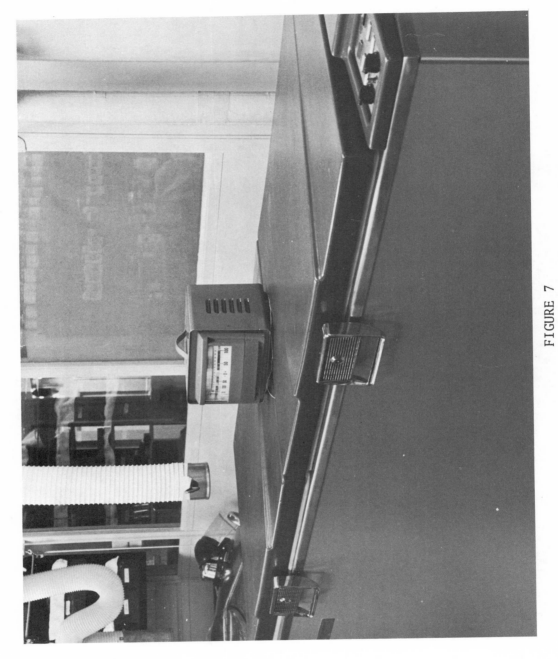

FIGURE 7

Union Carbide LR-1000 Liquid Nitrogen Refrigerator

Figure 8 depicts part of the inventory control system for storing the vials. The system consists of 52 racks. Each rack has 10 trays spaced 2 inches apart. Each tray contains a cardboard carton with dividers for 25 1-inch diameter vials. Record sheets are designed to coincide with the spaces allotted for storage of the organisms. Each set of organisms has its own numbering system. Storage space is assigned in proportion to the number of each group of organisms to be stored.

LABELING

All our vials are labeled with the UC (Upjohn Culture) number and the date of preparation. The following procedure renders the label indestructible by moisture on the vial as thawing occurs. The surface of the glass or of the cap to be labeled is coated with finger nail polish (any color--we prefer white--or brand may be used). It is desirable to coat the surface at least 30 minutes before marking to insure a hardened coat. Labeling is done with a Sanford's Fine Point Marker Sharpie No. 3000. Material so labeled may be repeatedly stored and retrieved from the liquid nitrogen refrigerator without destruction of the label. The Sharpie marker is also used to label the canes and cardboard boxes. It is not necessary to coat the aluminum or cardboard surfaces with nail polish. No moisture effect has been detected with these materials.

STORAGE, RETRIEVAL, AND VIABILITY

Storage

Unicellular bacterial cultures have been stored in the gas phase of liquid nitrogen in the LR-640C liquid nitrogen refrigerator since 1963. Filamentous bacteria (actinomycetes), fungi (including yeasts), and a few algae, protozoa, and phages have been stored in the gas phase in the LR-1000 liquid nitrogen refrigerator since 1971. The vials containing the material to be frozen are placed directly in the appropriate compartment in the refrigeration units. Vial caps should be sufficiently loose to permit a gaseous exchange. Vials with caps tight will fracture when frozen. The storage temperature ranges from -19 C at the liquid level to -75 C at the top of the storage unit. "Fog," which results from warm humid room air contracting the cold nitrogen gas when the lid is opened, is removed by an in-house ventilating system pulling a vacuum. This permits clear viewing of the contents. The "fog" is shown in Figure 9.

Retrieval

Frozen Suspensions. Vials are removed from the liquid nitrogen refrigerator, caps are loosened slightly, and the vials allowed to thaw at room temperature for 15 to 20 minutes. The aliquot needed for inoculation is pipetted onto the slant, or plate, or into the broth medium of choice and the vial is returned to the liquid nitrogen refrigerator.

FIGURE 8

Part of Inventory Control System for Storing Vials with Agar Plugs

FIGURE 9

"Fog" Which Results From Warm Humid Room Air Contacting Cold Nitrogen Gas When Liquid Nitrogen Refrigerator Lid is Opened

Each vial may be so treated until the contents are exhausted. We use the Falcon plastic disposable pipettes (Pipet No. 7506; 1 ml Serological; 1 in 1/100 ml with plug; sterile disposable) for dispensing liquid into and removing it from the 1/2 dram vials.

Frozen Agar Plugs. One straw is removed from the vial and placed in a petri plate. After a few minutes' thawing a plug is removed. The plug may be seeded on or into the desired medium or it may be fragmented in 2-3 ml sterile distilled water and aliquots seeded in or on the medium or media of choice. The straw with the remaining plugs is returned to the vial and refrozen. Straws are manipulated by means of a tapered device inserted into the slanted end of the straw. Ten ml pipettes (plastic or glass) are satisfactory. We are exploring the tooling of aluminum or stainless steel pencil-like probes, which may be sterilized by disinfectants, by autoclaving, or by dry heat, as more economical and space saving.

Viability

To date we have not had a viability problem. Unicellular bacteria have been used for seeding, for *in vitro* assays, for replacement of cultures used for specific purposes in various assay areas throughout the Company, and for specific fermentations in research. Frozen agar plugs have given reliable results when used for fermentation studies, for growth of assay organisms, and for miscellaneous research projects.

HAZARDS

Proper warnings should be posted. Hazards in the use of liquid nitrogen to be considered are:

1. Frostbite. Low temperature can and will cause frostbite if exposure is prolonged. Liquid nitrogen splash is not usually dangerous. *Touching cold metallic substances is extremely dangerous*, especially is the skin is wet and freezes to the metal surface. Clothing should cover the skin and face shields should be worn. We have found the use of 100% white cotton gloves ideal for handling frozen vials. These gloves, worn double, are light enough for good handling of small items and warm enough for handling several specimens at one time.

2. Explosion. As liquid nitrogen vaporizes, high pressure results from the expansion characteristic of the liquid. Rubber tubing should never be used to transfer liquid nitrogen. The rubber tubing becomes brittle and explodes like shrapnel. The same hazards exists when improperly sealed ampoules are removed from liquid nitrogen. (We have chosen gas-phase storage to overcome this hazard.)

3. _Anoxia_. Good ventilation is important. Normal room oxygen is 21%. If the level of oxygen in a room is displaced by nitrogen gas to 15% or less, dizziness and/or fainting may occur. Cold nitrogen gas goes to the floor where the fainting individual will have an even lower supply of oxygen.

4. _Fire_. Nitrogen gas is nonexplosive in terms of fire. Oxygen will condense on transfer lines of aluminum or copper. The hose will have a wet appearance and is a potential fire hazard. This is very minor hazard in most circumstances.

ADVANTAGES

Advantages of liquid nitrogen for preservation of microorganisms have been emphasized by Swoager[2] and Hossack[3]. Economical and reliable methods for preservation and reconstitution are primary concerns for individuals responsible for stock culture collections. Methodology for management of microbial preservation with minimal manpower is possible with cryogenic equipment--specifically gas-phase liquid nitrogen refrigerated storage tanks. Equipment and supplies, suitable for a culture collection of any size, are obtainable from a number of manufacturers. Electrical power needed to operate equipment with automatic controls is minimal. If a power shortage occurs, manual controls may be used. The cost of liquid nitrogen will vary according to the distance it must be transported and the adequacy of the filling system. The cost is far less than the salary of a good technician plus the cost of supplies needed for other methods of preservation.

SUMMARY

Gas-phase liquid nitrogen refrigerated storage is recommended for long-term preservation of microorganisms and especially for those organisms difficult to preserve by any other method. The procedures detailed are recommended for the maintenance of stable inoculum, for the preservation of genetic properties, for the ease of storage and retrieval of a large number or organisms by minimum personnel, and for the elimination of the variables that are unavoidable with the use of more tedious methods of stock culture maintenance.

---

[2]Swoager, W. C. (1972) Preservation of microorganisms by liquid nitrogen refrigeration. American Laboratory _4_:No. 12, pp. 45-52.

[3]Hossack, D. J. N. (1972) Liquid nitrogen frozen inoculua--a year's experience. Proc. Soc. Analyt. Chem. _9_:No. 2, pp. 36-38.

ACKNOWLEDGEMENT

I am indebted to many individuals for assistance in developing successful methods for storing microbes in liquid nitrogen refrigerated tanks.

PREPARATION, STORAGE AND RECOVERY OF
FREE-LIVING, NONENCYSTING CILIATED PROTOZOA

Ellen M. Simon, University of Illinois and
Miriam Flacks, University of California at Santa Barbara

The methods described here are used in laboratories maintaining specialized culture collections of ciliated protozoa, eukaryotic cells which have not been reported to survive freeze-drying. Approximately 340 stocks of *Tetrahymena pyriformis* are now stored in liquid nitrogen (LN) at the University of Illinois, the University of California at Santa Barbara, and the American Type Culture Collection, and 50 stocks of *Paramecium aurelia* at Indiana University. Reasonable recovery averaging 1-10% of nonencysting *Ciliata* representing 4 genera has been obtained[6].

Critical factors determining survival include the physiological condition of the cells, the presence of dimethyl sulfoxide (DMSO), two-step cooling with a carefully controlled rate between -15 C and the end of step one, minimizing warming of frozen samples being put into storage, rapid thawing, and dilution to reduce the concentration of DMSO and/or the effect of osmotic shock[3-6,9].

Adapting *Tetrahymena pyriformis* to abnormally high concentrations of NaCl, adding $MgCl_2$ to suspensions of cells, or adding NaCl, $MgCl_2$[9], sucrose or inositol[7] to diluents used at thawing gave inconsistent results. This behavior suggests that the survival of individual cells is governed by the interaction of many variables.

Multiple samples of each of 44 stocks of axenically-grown *T. pyriformis*, syngen 1 (syngen = genetically isolated biological species), were thawed following storage for five years. Recovery was as high as that from control samples thawed before storage; 94% of the samples were positive and the breeding performance, with two exceptions, was equal to, or better than, that of the same stocks kept at 15 C[7].

I. GENERAL PROCEDURES

A. Preparation of Cultures

Ciliated protozoa, many of which do not encyst, are cultivated in two basic kinds of media: (a) nonaxenic infusions in which bacteria or other small microorganisms serve as food (usually monoxenic - inoculated with a single species of bacteria) and (b) axenic broths. Some organisms grow more readily in one type of medium, others thrive in both types.

If a choice can be made, monoxenic culture has the advantage of not requiring strict asepsis. However, nonaxenic cultures to be frozen should be washed to eliminate any troublesome contaminants. Using a micropipette with an inside diameter slightly larger than the width of a cell, single cells are transferred quickly through five depressions in glass slides containing wash fluid, *e.g.* unbacterized medium or sterile Dryl's solution, allowing enough time for the animals to swim around in each depression. They are left in the fifth depression for four to five hours to permit digestion and elimination of ingested microorganisms, and then taken through five more transfers[10]. Axenic cultures can be established by using sterile media and equipment.

Ciliates which tolerate exposure to antibiotics may be rendered axenic by the following method: using aseptic technic and UV- or dry heat-sterilized equipment, add three drops of a healthy CA* culture with a Pasteur pipette to 0.5 ml of antibiotic solution (1 gm dihydrostreptomycin and 1,500,000 units of penicillin G in 50 ml sterile distilled water) in a depression slide. Allow to stand 30-45 minutes; transfer suspension to 5-10 ml of axenic medium. In a few days check for growth and contamination[1].

Cultures to be frozen are usually grown in Erlenmeyer flasks or in screw cap test tubes on a tube rotator, concentrated by centrifugation, mixed with DMSO, and equilibrated. For details see Section II.

B. Cooling and Storing

Starting at ambient temperature, samples are cooled in biological freezers at controlled rates to -40 C $\pm$ 15, then plunged into LN. After a minimum of 10 minutes, control samples are thawed and others are placed as quickly as possible in prechilled canes which are kept in a dewar containing some LN until filled. The cold ampules and canes are conveniently handled with the forceps shown in Figure 1.

C. Thawing

When sealed glass ampules are to be thawed, face and hands should be protected against possible explosion. This danger is reduced if the ampules have been stored in vapor over LN. If submerged in LN, placing the samples in vapor over LN in a small container for several minutes before thawing allows any LN inside improperly sealed ampules to escape. Explosions may be avoided by leaving glass ampules unsealed or by using plastic tubes.

Rapid thawing by plunging into a 20-40 C water bath results in greater survival than slower rewarming. The addition of sucrose, inositol, or sodium chloride to the diluent used at thawing *sometimes* further increases survival. However, these compounds, as well as DMSO,

---

*See section on media and solutions.

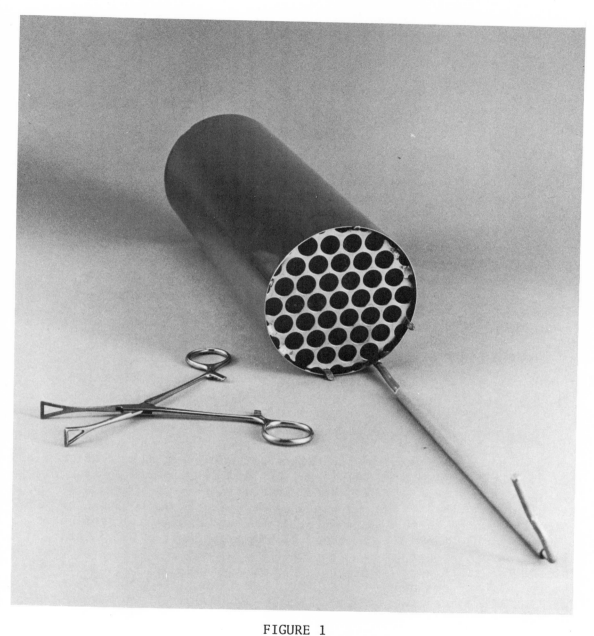

FIGURE 1

Canister from LD-30 LN Container with Perforated Metal Top and Triangular-tipped Obstetrical Forceps for Handling Ampules.

are toxic and must be diluted soon after thawing to below the following concentrations: DMSO - 2%, inositol - 1%, sucrose - 0.1 M, NaCl - 0.05M. If ampules are not sealed, diluents may be added quickly to intact ampules with an Aupette (Clay-Adams, Inc., New York, N.Y.). Bacterial contamination of axenic samples of ciliates unaffected by penicillin and streptomycin sulfate can be essentially eliminated by including these antibiotics in the diluting medium (250 units or ug/ml).

D. Examination for survivors (For *Paramecium aurelia* see II. C.)

Glass culture tubes into which thawed samples may be transferred during dilution (or in which preliminary experiment samples are frozen) are observed directly with a dissecting microscope by placing the tube horizontally, or nearly so, on the stage. Criteria for survival include motility and the capacity of the population to increase or of individual survivors to form clones. Motile cells may or may not be observed shortly after thawing. Healthy cells have been found 72 hours later in samples which appeared to contain no live cells at 24 or 48 hours.

Electronic counting is unsatisfactory because dead intact cells may be present. For a quick estimation of survival we use the following method. At a constant magnification the number of motile cells per field, or fraction thereof, is estimated and multiplied appropriately. The results for the entire sample are recorded on an arbitrary scale which is roughly logarithmic (1=1-10, 2=10-40, 3=40-150, 4=150-500, 5=500-1500, 6=1500-5000, 7=>5000). For more accurate counts of viable cells 0.1 ml of a culture or of a frozen-thawed sample may be diluted to include 30-100 cells which are enumerated by either of the following methods: (a) 0.05 ml is distributed in very small drops on a microscope slide[3] or (b) 0.1 ml is divided into three slide depressions, the motile cells are removed individually with a micropipette while counting.

These methods yield immediate results but no information on the cloneforming capabilities of individual cells. Estimates of the number of such *T. pyriformis* cells utilizing the Poisson distribution have been made by two methods. (1) Serial 10-fold dilutions made from samples immediately after thawing were divided into as many aliquots as convenient. Antibiotic medium was used for axenic samples. After incubation for two to three days, one dilution should yield some aliquots with, and some without, survivors. (2) Heaf and Lee[2] diluted their cultures to a concentration of 50 to 400 organisms/l. While the contents of the flask were being stirred magnetically, 0.5 ml samples were pipetted with a Cornwall syringe continuous pipetting outfit (Becton Dickinson Ltd.) into the cups of sterile 80-hole perspex hemagglutination trays. Cups containing viable organisms were counted with the naked eye after 6 days at 28 C.

II. DETAILED PROCEDURES

A. *Tetrahymena Pyriformis*

A study with one monoxenically-grown stock of the survival of cells taken at two-to six-hour intervals from large cultures suggested that maximum recovery occurred only when the culture was at the transition from logarithmic (ultradian) to stationary (infradian) plus or minus three or four hours[6] (see Figure 2). The difficulty in achieving this condition at a convenient time can be reduced by pooling two or more cultures started at different times, with different initial concentrations of cells, incubated with or without shaking or at different temperatures (usually between 20 and 27 C; some stocks also grow well at 30 C).

Inoculate cultures to be processed with early infradian cells (volume 2 to 5% of the total).

1. Axenically-grown animals have been stored successfully by the following procedures. Grow seed and final cultures in tetrahymena broth.

    a. Flask cultures at the critical stage of growth contain an average of 2 to 3 x 10^5 cells/ml. They should be handled aseptically through the entire process. Concentrate approximately 25 times by centrifuging at 1300 to 1400 rpm for 2.5 minutes and decanting immediately. Mix with an equal volume of 20% DMSO in sterile demineralized H_2O or in peptone broth.

    Dispense 0.1 or 0.2 ml volumes in glass ampules during a period of equilibration at room temperature lasting approximately 30 minutes.*

    Cool with the rate accelerating during the first 3-4 minutes from 0 to between 6.5 and 10 deg/min to the plateau. Adjust rate to 1.5-2 deg/min to -25 or 30 C. Maintain final temperature until total cooling time is 20 minutes.

    Plunge into LN. After a minimum of 10 minutes, store in the frozen state or thaw individual samples with agitation in a water bath at 30 C $\pm$ 7 for 35-40 seconds.

---

*Samples (0.2-ml) of some stocks were equilibrated at 26 or 35 C and cooled for 20 minutes in a household upright freezer in an ethylene glycol-dry ice slush at -30 to -35 C or in a Nitro Freeze apparatus at -23 C, then placed in the vapor area of a LN tank. Thawed samples were diluted serially at 2-10 minute intervals with 0.2, 0.3, 0.5, and 1.0 ml of sterile medium[2,7].

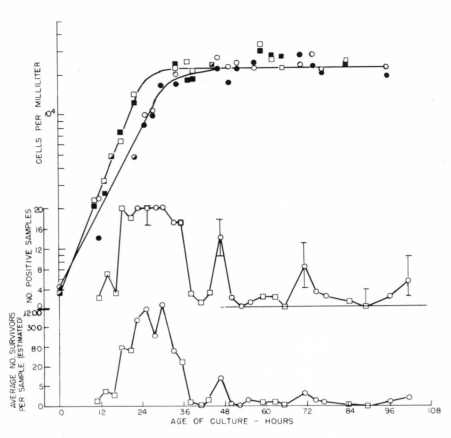

FIGURE 2

Relationship Between Growth Curve and Survival of *T. pyriformis*, 2 I, After Exposure to -196 C. Flasks A and B (Circles) Were Inoculated 12 Hours Before C and D (Squares); Monoxenic Growth Medium. Data on Survival of Cells Taken From Duplicate Flasks Were Pooled. Bars Indicate 95% Confidence Limits. (Published by Permission of International Institute of Refrigeration).

Quickly open and dilute by one of the following methods: (a) add 0.5ml tetrahymena broth or proteose peptone broth; within 2-5 min transfer to culture tube and add a second 0.5 ml; or (b) add 0.3 ml of a 2% solution of inositol in medium, then two 0.3-0.4 ml additions of plain medium at 2-3 minutes intervals.

b. Differs from a as follows:

Grow in PPY medium in screw cap tubes on a tube rotator (60 rpm) at 25 C for 1-3 days depending on rate of growth of the strain. Immediately after removing from the rotor, centrifuge at 310 g; remove 75% of the supernatant and add 20% DMSO. This concentration (*ca* 3.5 x 10^5 cells/ml) was best for a wild type syngen 1 strain; mutant strains required more concentrated suspensions.

Dispense 0.2 ml volumes in plastic tubes and tighten caps.

During cooling, after a 1-2 minute freezing plateau, use "override" switch to lower temperature at 2 deg/min. With plastic tubes the rate of cooling increases to 3-5 deg/min between approximately -15 and -20 C; thereafter it can be maintained at 2 deg/min to -40 C.

Following at least 24 hrs. in LN thaw samples in a 30 C water bath with agitation for 45 seconds, add 1.0 ml of PPY (+ penicillin 250 units/ml and streptomycin sulfate 250 ug/ml). Recap and incubate for 1-3 days.

2. Monoxenic cultures contain an average of 2-5 x 10^4 cells/ml.

Concentrate 35-70 times.

Add DMSO to a final concentration of 10%.

Dispense 0.1 or 0.2 ml volumes in 0.7-ml ampules during a period of equilibration at room temperature lasting approximately 30 minutes. Leave unsealed.

Cool at a rate of 10-15 deg/min to plateau. Continue at 2-2.5 deg/min to -40 or -50 C.

Plunge into LN and store or thaw as for axenic cultures. More cells have been recovered from some stocks thawed at 23 C than at 37 C and *vice versa*[6].

Flood samples with 0.5 ml exhausted CA immediately before or during thawing, transfer to culture tubes and add another 0.5 ml.

B. Other *Tetrahymena* spp., *Glaucoma*, and *Colpidium*

In limited experiments *Tetrahymena vorax*, *Tetrahymena setifera*, *Glaucoma chattoni*, and *Copidium campylum* grown monoxenically and treated as described in A.2. were recovered. Some stocks of *Glaucoma*, more sensitive to DMSO, have not yet survived exposure to -196 C. The addition of 0.05 M sucrose to thawing diluent may increase the survival of *T vorax* and *Glaucoma*.

Tetrahymena limacis and *Tetrahymena rostrata* grown anenically in skim milk medium and cooled as in A.2. were recovered[6].

C. *Paramecium aurelia* grown monoxenically

1. Syngens 1 and 4. Grow at 1 fission per day (by doubling the volume with fresh CA daily) for 4-6 days.

    Concentrate to 200,000 cells/ml.

    Mix with an equal volume of 15% DMSO in CA medium.

    Dispense 0.2-ml volumes in ampules during a period of equilibration lasting 30-60 minutes.

    Cool at 5-10 deg/min to plateau, adjust rate to 2-3 deg/min and continue to -40 C.

    Plunge into LN. After 30 minutes store, or thaw by dropping into a 30-35 C water bath. After a few seconds, open and flood with 0.75-1.0 ml Dryl's solution at the same temperature.

    Distribute in several depressions, examine, and add 0.25 ml CA medium to syngen 1 or BL medium to syngen 4. Count again after 18 hours. Unlike *Tetrahymena* many paramecia which survive freezing and thawing are permanently injured[9].

2. Syngens 2 and 14. One stock of each has been recovered when treated by a method differing in the following respects from that in C.1. CA flask cultures were started with 1.5-2% inoculum, incubated four days at 23 or 30 C, and concentrated by centrifuging for four minutes. The 15% DMSO solution was made in exhausted CA. Cooling below the plateau was at 1 deg/min to -50 C. Supercooling was minimized by use of the "override" switch. As quickly as possible 0.8 ml of exhausted CA was added and the sample thawed with agitation in a 23 C water bath. Fresh CA (0.7-1.0 ml) completed the diluting process. "Negative" samples in tubes were reexamined at 48 and 72 hours[4].

D. *Cyrtilophosis mucicola*,

C. mucicola a cyst-forming ciliate, was included in one experiment for comparison. A suspension of cysts without a cryoprotectant was air-dried on filter paper, then placed in culture tubes. The cysts survived two-step cooling to -196 C[9].

III. EQUIPMENT AND SUPPLIES

A. Biological Freezing System; BF-3 or BF-4 (Union Carbide Corp., Linde Division) with Honeywell Electronik 18 Temperature Recorder (ambient to -100 C) or Leeds & Northrup Speedomax H, Model S, strip chart recorder.

B. LN source for freezing system: (University of Illinois) LD-31 Linde LN container fitted with the dispensing unit shown in Figure 3. Large pressurized tanks are impracticable for us since cheap LN is available two blocks from our laboratory. This tank doubles as a conveyor of LN for our storage containers and as a low pressure source of LN for use with the BF-3 Freezer. Adequate pressure builds up over a weekend or can be hastened by admitting pressurized air to the tank. A safety valve set at 10 psi is attached.

(University of California) Union Carbide UC-50 tank fitted with valve, pressure gage and connector.

C. LD-30 Linde LN containers for storage. Locating a particular stock has been facilitated by placing perforated metal plates in the top of each canister (see Figure 1). Circles with four "ears" were cut from purchased perforated sheet metal and the perforations were enlarged to accommodate canes containing ampules.

D. Microscope: B & L Stereozoom dissecting with substage mirror and transmitted light.

E. Glassware, etc.

Culture tubes: 10 x 75-mm round bottom, for cooling samples in preliminary experiments. Thawed, diluted samples may be left in these tubes for examination and, unlike ampules, they can be reused many times.

Glass ampules: Cryules (Wheaton, Millville, N.J.). 1.0-ml (or 0.7-ml) for those to be sealed; 0.7-ml for those left unsealed. Six of the latter size fit in a 6-ampule cane if placed as close as possible.

Plastic tubes: A/S NUNC, screw cap, 38 x 12.5 (4-shore USA, P.O. Box 264, La Jolla, Calif. and several suppliers of cryogenic equipment).

Pyrex glass slides with three depressions - capacity approximately 1 ml each.

Aluminum canes: 6-ampule (Shur-Bend Mfg. Co., Minneapolis, Minn.).

Serological pipets: 1-ml with long tips for filling ampules.

Forceps with triangular tips (see Figure 1) for handling cold ampules and canes.

FIGURE 3

LD-31 LN Container With Closure Permitting Withdrawal Into BF-3
Freezing System or Into Other Containers.

Sanford's "Sharpie" pen for labeling ampules and canes.

F. Media and Solutions

1. Monoxenic growth media (infusion of plant material supports growth of bacteria which are the principal food for the protozoa).

   a. CA (Cerophyl Infusion). Of several cereal powders distributed under the trade name Cerophyl (Cerophyl Corporation, Kansas City, Mo.) that made from rye is most suitable. Prepare by boiling for five minutes 1.5 or 2.5 gm/l of distilled water, filter, add 0.5 gm $Na_2HPO \cdot 12\ H_2O$/l, and autoclave. (The phosphate buffer is not required for cultivation of *Tetrahymena pyriformis*.) Inoculate with *Enterobacter (Aerobacter) aerogenes* and incubate overnight at 25-35 C before adding ciliates[10]. If refrigerated, the bacterized medium may be used for several days.

   b. BL (Baked Lettuce Powder Infusion). Rinse lettuce in tap and distilled water; spread leaves in single layer and bake until light brown at 180 C or bake entire head for 48 hours at 110 C. Discard unbrowned portions, grind to a coarse powder. Add 1.0-1.5 gm/l or redistilled water and heat to boiling, filter, add enough $CaCO_3$ to saturate and autoclave. Incubate with *E. aerogenes*, as above, adjust pH to approximately 7 with supersaturated $Ca(OH)_2$ before inoculating with ciliates[10].

2. Axenic media (Some batches of proteose peptone are unsatisfactory.)

   a. Peptone broth. Difco proteose peptone - 1% in distilled H_2O. Autoclave.

   b. PPY broth. 20 gm Difco proteose peptone and 1 gm Difco yeast extract/1 H_2O Coarse filter and autoclave.

   c. Tetrahymena broth. 5 gm Difco Proteose peptone, 5 gm tryptone and 0.2 gm KH_2PO_4 in 1 liter distilled H_2O. Autoclave[3].

   d. Skim milk. Ten gm skim milk powder, 5 gm tryptone or Difco proteose peptone, 2.5 gm yeast extract. Adjust to pH 7 with NaOH. Autoclave.

3. Solutions used in thawing and subsequent dilution.

   a. Dryl's solution. To 945 ml of glass-distilled H_2O add 10 ml of 0.1 M (=1.380 gm/100 ml) $NaH_2PO_4 \cdot 1H_2O$, 10 ml of 0.1 M (=2.681 gm/100 ml) $Na_2HPO_4 \cdot 7H_2O$, 20 ml 0.1 M (=2.941 gm/100 ml) sodium citrate ($2H_2O$), and 15 ml 0.1 M (=1.470 gm/100 ml) $CaCl_2 \cdot 2H_2O$. To avoid precipitation, add $CaCl_2$ last[10].

b. Exhausted medium. The supernatant from CA cultures centrifuged in preparation for freezing is collected in a stainless steel beaker which is placed in a LN bath until completely frozen to kill remaining protozoa. Although not literally exhausted, the supply of food is greatly reduced. The thawed medium can be dispensed in culture tubes in quantities suitable for diluting single samples and stored frozen, *e.g.* at -20 C.

These studies were supported chiefly by the following grants: U.S. Public Health Service GM 07779 and National Science Foundation GB 23410 to D. L. Nanney, U.S. Public Health Service GM 15140-01 to T. M. Sonneborn and National Science Foundation GB 13207 to Eduardo Orias.

REFERENCES

1. Allen, S. L. Unpublished.

2. Heaf, D. P. and Lee, D. A viability assay for *Tetrahymena pyriformis*. J. Gen Microbiol. 68, 249-251, 1971.

3. Hwang, S-W., Davis, E. E., and Alexander, M. T. Freezing and viability of *Tetrahymena pyriformis* in dimethylsulfoxide. *Science 144*, 64-65 (1964).

4. Simon, E. M. *Paramecium aurelia*: recovery from -196°C. *Cryobiology 8*, 361-365 (1971).

5. Simon, E. M. Freezing and storage in liquid nitrogen of axenically and monoxenically cultivated *Tetrahymena pyriformis*. *Cryobiology 9*, 75-81 (1972).

6. Simon, E. M. The preservation of Ciliated Protozoa in Liquid Nitrogen. In press. Proceedings of symposium on "Freeze-drying of Biological Materials" held in Sapporo, Japan, October 7-10, 1973; auspices International Institute of Refrigeration.

7. Simon, E. M. Unpublished.

8. Simon, E. M. and Hwang, S-W. Tetrahymena: effect of freezing and subsequent thawing on breeding performance. *Science 155*, 694-696 (1967).

9. Simon, E. M. and Schneller, M. V. The preservation of Ciliated protozoa at low temperature. *Cryobiology 10*, 421-426 (1973).

10. Sonneborn, T. M. Methods in Paramecium Research. In "Methods in Cell Physics," vol. 4 (1970). Academic Press, New York. pp. 241-339.

PRESERVATION OF CULTURES IN ROUTINE
MICROBIOLOGICAL ASSAY OPERATIONS

Walter T. Sokolski

Infectious Diseases Research
The Upjohn Company

Microorganisms have been preserved in many ways for long periods of time. The maintenance of culture lines was the usual reason for preservation and the method used was satisfactory as long as some cells remained viable upon transfer. Most methods were time-consuming. With the recent availability of cryogenic materials and equipment, cultures could now be frozen and stored for long periods and with no detectable loss in viability. This paper is concerned with the freezing and storage of full-grown cultures for direct use as inoculum in microbiological assays.

The cryogenic freezing and long-term storage of suspensions of microorganisms used directly as inoculum in vitamin and antibiotic assays has been routine in our laboratories for over 10 years. Generally, suspensions are prepared in the same way as for assay with fresh inocula, distributed in tubes, quick frozen and stored. Adjustments of suspensions are usually done before freezing. Each batch is checked for dose-response of a control solution before putting into routine use.

Cryogenic storage proved to have many advantages over the old daily-transfer method. There is less contamination, less changes-in-culture, and less requirement for technical manpower. Once the suspensions have been frozen, a sample is simply removed from the freezer and used as if it were a chemical reagent. The most important advantage is a reduction in day-to-day variation in the assays. The control laboratory of Upjohn Limited in England has used our frozen cultures for some time. Hossack[1] summarized a year's experience and listed the advantages as follows:

    Increased reliability
    Time saving
    Constant availability of inocula
    Reduced repeat test
    Increased confidence between laboratories
    Reduced need for a specialist microbiologist

CONTROL PROGRAM

A cryogenic program was started in the control laboratories in 1962. We planned to standardize assay organism suspension for our use and for other laboratories in the company. Because of our limited storage space, we initially stored suspensions under liquid nitrogen for which we used sealed glass ampoules. Ampoules are the containers of choice for organism suspensions where large numbers are used and storage space is limited.

Our early studies were with small numbers of ampoules stored in 25-liter liquid nitrogen refrigerators. We gradually expanded to larger storage tanks, a semi-automated ampoule-sealer (Kahlenberg-Globe, Sarasota, Fla.) and an ampoule labelling machine. In our present operation, we label the ampoules with waterproof ink and then bake the label in for five minutes at 100 C. The ampoules are then sterilized in the autoclave, filled, sealed, and fast-frozen.

In the freezing process, the ampoules are placed in canes, the canes in canister, and each canister is dipped into liquid nitrogen in a 25-liter or larger tank. The warm canister causes the nitrogen to boil. We now use a 35-liter tank (Cryenco, Denver, Colo.) with a lazy-susan turntable and an off-center opening. We drop the canister in the opening and immediately turn the lazy-Susan so that the boiling does not occur at the opening and possibly spill over.

Hazards and Precautions. The type of ampoule used should be suitable for use in liquid nitrogen (Borosilicate glass for LN_2 use, Kimble or Wheaton). These should be sterilized in an autoclave or by ethylene oxide and not by dry heat. We experienced more breakage of ampoules sterilized in the oven. If ampoules are stored under liquid nitrogen, when one is being removed, it is advisable to set the ampoule in the gas phase for a time before removing it from the nitrogen container. This is a safety precaution for potential leaker-explosions where the liquid in the "leaker" is allowed time to evaporate off.

Vitamin Assays. Our first studies were with vitamin B_{12} and pyridoxine assays. In our routine assays, the vitamin B_{12} organism was *Lactobacillus leichmanii* UC 240, usually grown in tomato juice broth, centrifuged, washed, and suspended in saline.[2] Frozen saline suspensions (fast frozen) however required longer incubation time for the same assay response as the fresh (Figure 1). Note in Figure 1 also that the fast frozen was better than the slow-frozen. When the cells were suspended in vitamin B_{12} assay medium and then fast-frozen, the assay response was parallel to one with a fresh suspension. All frozen suspensions were fast-thawed by agitation in a 40 C water bath. The viable cell recovery on the frozen suspension in Figure 2 was 100 per cent.[3]

The pyridoxine assay was usually done with *Saccharomyces carlsbergensis* UC 4920 (Fleishman 4228) cells grown on Sabouraud's maltose agar, washed, and suspended in saline.[4] Freezing the cells in saline did not

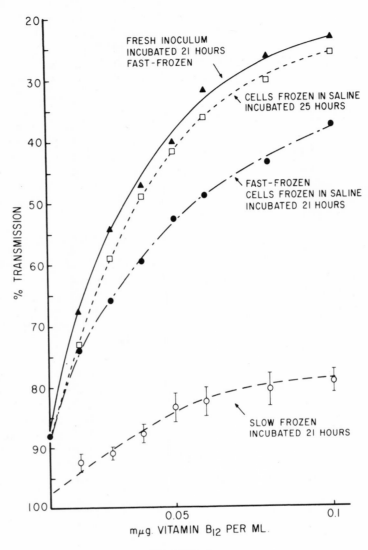

FIGURE 1

Freezing and Assay Conditions for *Lactobacillus Leichmanii* UC 240. Fast-Frozen is by Direct Immersion in Liquid Nitrogen. Slow-Frozen is With Lowering Temperature 1 deg/min to -40 C.

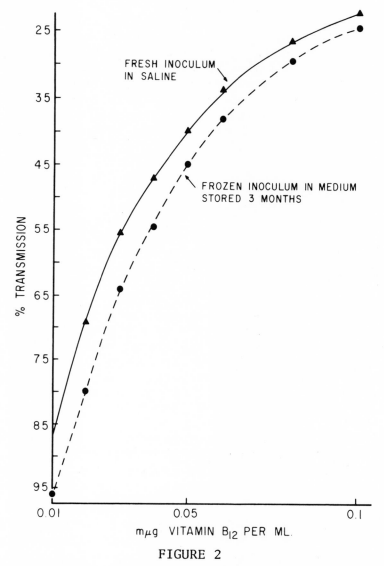

FIGURE 2

Dose-Response Curves for *Lactobacillus Leichmanii* UC 240.

work for the assay (Figure 3). Once again, the best condition was fast freezing and fast thawing of the cell suspension in the basal assay medium.

Antibiotic Assay. A commonly-used antibiotic medium No. 1.[5] A suspension of growth was prepared in antibiotic medium No. 3 (broth), standardized,[6] distributed to ampoules and frozen as described above. The frozen suspension was tested in lincomycin agar diffusion assays and compared with 2 aliquots of the same cell suspension stored at 4 C.[7] The inoculum rates for the assay agar was 0.7% for the frozen and for 1 aliquot of unfrozen suspension. For the 2nd unfrozen aliquot, the inoculum was 0.7% and gradually increased to 1.4% until the end of experiment (70 days). Each assay was run with lincomycin concentrations at 0.5, 1.0, 2.0, 4.0, 8.0 and 16.0 µg/ml. The slopes of the standard curves were calculated and plotted as shown in Figure 4. Note that the slope with frozen inoculum was fairly constant throughout while the slopes of the unfrozen inoculum changed after 30 days.

Summary of Organisms. A total of 41 strains of bacteria, yeast, and molds used in analytical tests are prepared and stored in glass ampoules for use as direct inoculum in assays in our control laboratories and are also shipped to other laboratories in the United States and in many foreign countries. Of these, there are 29 different species and 20 different genera including two of yeast and three of molds.

Freezing and Storing of Reference Standards and Media. Stock solutions of all vitamin and antibiotic standard preparations are stored frozen, ready for use when needed. Most solutions are distributed in 1 ml to 20 ml screw cap vials and are stored in the gas phase in liquid nitrogen refrigerators. This system is convenient not only for the saving of material and analyst time but also for better standardization between laboratories. A total of 23 standard reference solutions of vitamins and antibiotics are stored frozen in our control laboratories.

Media for 8 vitamin assays are prepared in large quantities at 2x or 4x concentrations. These are stored in 500 ml Wheaton tissue culture bottles (Vitro 400, Wheaton Glass Co.) with 100 ml to 250 ml per bottle. This is also a material and time-saver.

PLASTIC TUBES

A markedly different system for cryogenic storage was designed for our antibiotics bioassay laboratory in Infectious Diseases. Instead of glass ampoules, we store our test organism suspensions in polypropylene snap-cap tubes (Falcon Plantic, cat. 2005, 2063, 2006, 2059). These are better for an operation where assay systems are changed frequently. We may put up as few as five and as many as 500 tubes of a culture.

Most of the organisms used for assay are bacteria. Since they can stand more stress in temperature changes than other organisms, viable recovery is good whether suspensions are stored in glass or plastic

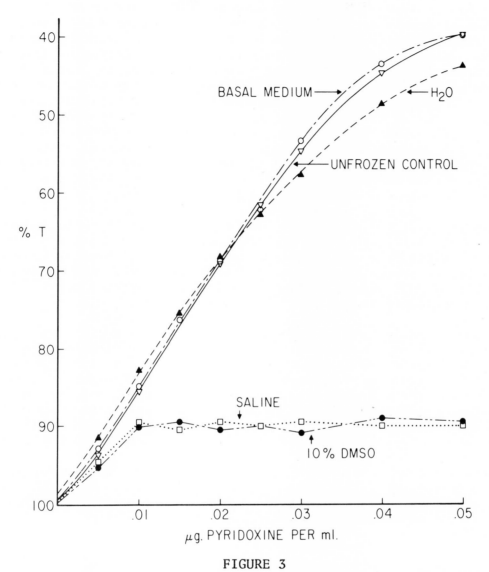

FIGURE 3

Pyridoxine Dose-Response Curves With *Saccharomyces Carlsbergensis* Frozen in Various Media.

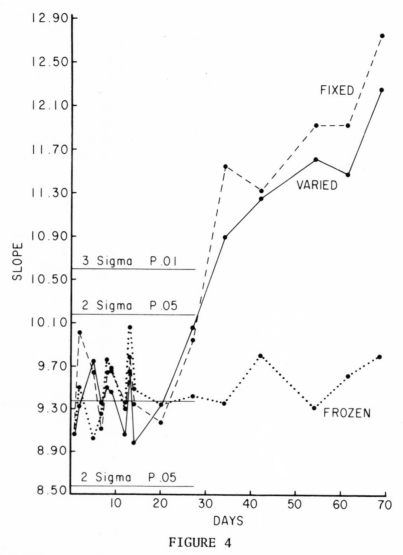

FIGURE 4

Slope Variations With *S. Lutea* Suspensions in Lincomycin Assays.

tubes. Our experiences with cultures of *Escherichia coli* and *Staphylococcus aureus* indicate that the suspension medium is more important for viable recovery than the rate of freezing and thawing.

Standard Procedure. A typical assay organism is *S. aureus* UC 80 (FDA 209P). The usual procedure is to culture the organism in one or more 500 ml Erlenmeyer flasks with 100 ml trypticase soy broth in each flask. The flasks are incubated on a shaker at 250 RPM and 37 C. for 18-20 hours. The cultures are pooled and the suspension is stirred on a magnetic stirring apparatus. The culture is distributed to 6 ml polypropylene tubes (Falcon Plastic cat. 2005, 2063). with 1 ml per tube. Each tube is code-labelled with a nonwashable felt pen. The tubes are placed in racks and frozen either by direct immersion in liquid nitrogen or by standing in the gas phase above the liquid. The tubes are stored in the gas phase of a liquid nitrogen refrigerator where the temperature range is between -75 to -196 C. On the day of assay, a tube is removed and the suspension thawed by adding 3 ml of a nutrient broth. The contents of the tube is added to 1 liter of agar at 48 C. The seeded agar is dispensed to assay plates or trays.

Some assay suspensions, such as the yeasts, are better when prepared from growth on solid medium. A typical preparation, *Saccharomyces cerevisiae* UC 1337, is grown on the following agar in gm per liter:monopotassium phosphate 5, dextrose 30, yeast extract 7, agar 15 (pH 5.3). Roux bottles, each 250 ml agar, are inoculated and incubated for 48 hours at 28 C. The growth is washed off the surface of the agar with broth of the same composition (less agar) and glass beads. About 40 to 50 ml suspension is made from each bottle. The suspension is distributed to polypropylene tubes with 1 ml per tube, frozen, and stored for assays as described above.

Fungi are frozen and stored as spore suspensions in physiological saline. A typical organism, *Glomerella ungulata* UC 1211, is grown for one week or more at 28 C on malt extract agar consisting of 2% malt extract, 2% dextrose, 0.1% peptone, and 1.5% agar (pH 5.4) in 150 x 25 mm plates. Twenty ml saline is added to each plate and the spores are scraped off the growth mat with a glass "hockey stick." The suspension is filtered through glass wool to remove any clumps of mycelium. The suspension is distributed to tubes and frozen as described above.

Another organism which is prepared differently from the methods described above is *Mycobacterium avium* UC 159. One or more 100 ml shake cultures in 500 ml Erlenmeyer flasks are incubated shaken at 37 C for 24 hours. The broth medium consists of the following in gm per liter: cerelose 10, peptone 4, beef extract 4, yeast extract 1, sodium chloride 2.5, and polysorbate (80) 10. The cultures are pooled and centrifuged at 6000 RPM (5860 x g). The supernatant liquid is discarded and the cells are suspended with 0.1 original culture volume in brain heart infusion broth. The suspension is distributed to tubes, frozen, and stored as above. In this procedure, the cells are concentrated 10-fold over the original 24-hour culture. Another reason for centrifuging the cells is to eliminate polysorbate 80 from the assay medium. Polysorbate 80 interfered with the agar attaching to the plastic surface of the assay plate. The agar tended to slide off the plate when held at an angle.

Summary of Organisms. A total of 44 strains of bacteria yeast and molds used in analytical tests are prepared and stored in plastic tubes for use as direct inoculum in assays. Of these, there are 27 species of bacteria, 3 yeasts, and 3 molds.

Organisms Used for *In vivo* Tests. Approximately 80 pathogenic strains of bacteria, yeasts, molds, and protozoa have been stored at temperatures of -40 C or below for 20 or more years.[8] This includes 3 strains of *Plasmodium berghii*, stored for 9 years in 5% dimethyl sulfoxide in rabbit blood. No evidence for loss of pathogenicity has ever been noted.

REFERENCES

1. Hossack, D. J. N. 1972. Liquid nitrogen frozen inocula - a year's experience. Proc. Soc. Analyt. Chem.: 36-38.

2. *The Pharmacopeia of the U.S.* XVIII. 1970. p 888.

3. Sokolski, W. T., Stapert, E. M. and Ferrer, E. B., 1964. Liquid nitrogen freezing in microbiological assay systems 1. Preservation of *Lactobacillus leichmanii* for direct use in vitamin B_{12} assay. *Appl. Microbiol.* 12:327-329.

4. Sokolski, W. T. and Stapert, E. M., 1964. Liquid nitrogen freezing in microbiological assay systems. III. The preservation of test organism and medium for direct use in the microbiological assay of pyridoxine. *Developments in Industrial Microbiology* 6:178-184.

5. Kirshbaum, A. and Arret, B., 1967. Outline of details for official microbiological assay of antibiotics. *J. Pharm. Sci.* 56:511-515.

6. Kirshbaum, A., Kramer, J. and Garth, M.A. 1962. Uniform preparation of microbial suspensions for antibiotic assays. *Antib. and Chemotherapy* 12:545-549.

7. Stapert, E. M. Sokolski, W. T. Kaneshiro, W. M. and Cole, R. J. 1964. Liquid nitrogen freezing in microbiological assay systems. II. Preservation of *Sarcina lutea* for antibiotic assays. *J. Bact.:* 532-533.

8. Lewis, C. Unpublished Data. The Upjohn Co., Kalamazoo, Michigan 49001.

CRYOPRESERVATION OF PROTOZOA AND HELMINTHS

Kenneth W. Walls

U.S. Department of Health, Education, and Welfare

Protozoa and helminths are giants in comparison with other microbial organisms such as the viruses, rickettsia, and bacteria. Furthermore, cryopreservation of the pathogenic species differs from that of the free-living species because not only must the organism *per se* be preserved, but in addition its characteristics of pathogenicity, virulence, and antigenic makeup must be maintained. Consequently, the problems of cryopreservation are unique and require special consideration.

As early as 1939, Coggeshall[8] reported the successful preservation of the malaria parasites by freezing in a mixture of dry ice and ethanol and then storing in dry ice. After 70 days he found the frozen *Plasmodium knowlesi* still infective for monkeys. This demonstration of the resistance of a protozoan to freezing and thawing paved the way for widespread studies on the factors and procedures for the cryopreservation of pathogenic protozoa and helminths of man.

Two excellent reviews have been prepared on this subject. The earliest is that of Diamond in 1961[15] in which he includes all of the reported species of protozoa which had been successfully frozen. He reviewed 70 reports covering 44 species of protozoa. In 1972 Dalgliesh[14] cited 34 additional reports with 29 species. It now becomes apparent that although there are marked differences in species resistance, it is likely that all protozoa will withstand cryopreservation if we only delineate the conditions necessary for their survival. Table I lists the genera of protozoa that have been studied to date. By far the most intensely studied genera are the plasmodia with 18 species studied by 17 investigators and the trypanosomes with 21 species studied by 11 investigators. A review of the reports of these investigations also reveals that these two genera have been the subjects of the greatest number of studies in which the protozoan was successfully frozen without the aid of a protectant and frequently without a controlled freezing rate. Although only speculative, it has been postulated that these organisms are particularly resistant because they are found in the blood or in the blood cells, and this protects them from the adverse effects of rapid freezing and thawing. The higher density of the blood or serum medium in which the organisms are found would tend to prevent drastic osmotic pressure changes during freezing and to decrease the rapid transport of water across the cell membrane.

TABLE I

GENERA OF PARASITIC PROTOZOA PRESERVED BY FREEZING*

GENUS	NO. OF SPECIES	NO. OF INVESTIGATORS
BABESIA	8	8
EIMERIA	11	3
ENTAMOEBA	2	2
GIARDIA	1	1
LEUCOCYTOZOON	1	1
PLASMODIUM	18	17
TOXOPLASMA	1	6
TRICHOMONAS	7	7
TRYPANOSOMA	21	11
MISCELLANEOUS	3	3

*Compiled in part from Diamond[15] and Dalgliesh[14]

It is interesting to note the paucity of reports for those protozoa which are found outside the blood: two reports for *Entamoeba*, one for *Giardia*, and six for *Toxoplasma*. Until recently, all attempts at cryopreservation of these organisms have been unsuccessful.

TECHNIQUES

Before individual organisms are discussed, a review of the techniques and reagents selected for specific purposes may be helpful.

Protectants: With the elucidation by Lovelock[28] of the action of glycerol in the protection against hemolysis during freezing and thawing, the use of bioprotectants has become almost standard in the cryopreservation of protozoa. Although many compounds have been tested[12,16], only two have been found useful - glycerol and dimethyl sulfoxide (DMSO).

By far the most commonly used agent is glycerol because of its relative availability and low cost. Practically every species of protozoan studied has been successfully frozen with glycerol[14,15,48]. Its use, however, has not been without drawbacks. Lovelock showed some time ago[28] that an equilibrium period was necessary for the glycerol to pass through the cell membranes in order to offer full protection. A number of studies have shown that excessive amounts of glycerol are injurious to the cells even when they are not subjected to freezing[27,35].

Dalgliesh[13] reported an interesting phenomenon that indicates the possible effect of glycerol on the host. When *Babesia bigemina* which had been preserved in glycerolized blood was inoculated intravenously into susceptible hosts, there was no response; but when it was inoculated subcutaneously, infections occurred in 100% of the hosts. On the other hand, when organisms that had been frozen without glycerol were thawed and injected, exactly the opposite was seen - intravenous inoculations were successful but subcutaneous inoculations failed. The authors had no explanation for this other than the toxicity glycerol exhibits on the host.

Although concentrations of glycerol as high as 20% (v/v) have been used successfully[2], most authors find concentrations of 7% - 10% to be optimum. Similarly, most authors dilute the glycerol in a buffered salt (pH 7.2-7.4), but Levine *et al.*[27] report that the protection of *Trichomonas fetus* is greater when the glycerol is diluted in water than when it is diluted in physiological saline.

In order to overcome these problems, many authors have used DMSO as an alternative[12,40,46]. As with glycerol, a variety of concentrations of DMSO have been suggested. Dalgliesh[12] suggests 4M (14.4%) as the optimum for *Babesia bigemina*, whereas Callow and Farrant[6] report 1.5M as the most effective for *Leishmania tropica*. In general, most investigators have found 1.5 - 3.0M (5%-10%) DMSO to be optimum.

Lumsden[29] reported that *Trichomonas vaginalis* does not survive freezing in the presence of glycerol but does exceptionally well with the addition of 10% DMSO. In this series of experiments, he used motility as a measure of survival and reported that 85% - 95% of the organisms survived 85 days in dry ice with DMSO but only a few survived the same period of time with glycerol. Organisms in the presence of glycerol immediately became distorted and lost motility, while those in DMSO showed no adverse effect. Although there was an immediate loss of activity with the addition of DMSO, this was only minor and no further loss occurred for the remainder of the experiment. Of even greater importance is the fact that there were no adverse effects of the DMSO seen upon subculture.

In some cases protective agents have not been used, and in others, they have actually been found harmful. Smith[45] showed that with *Toxoplasma gondii* he was able to freeze the culture at least 160 days when it was suspended in fetal calf serum with no additives. When glycerol or DMSO were added singly or in combination, the organisms would not survive 8 days storage in liquid nitrogen. The helminths have been preserved completely without protective agents[5,7,21], and many investigators have found protective agents unnecessary for freezing the plasmodia[8,22,23,32-34,42]. Herbert et al.[20] found that trypanosomes preserved in blood taken directly from the infected host and frozen instantaneously in liquid nitrogen had a higher percentage of survival than those treated with both glycerol and DMSO.

Rate of freezing and thawing: Most investigators have reported that slow freezing is necessary for maximum survival of the organisms. Dalgliesh[14] and others[10,13,18] suggest decreasing the temperature by 1 deg/min between 0 C and the temperature of storage. Others have suggested decreasing the temperature 1 or 2 deg/min from 5 C to below -30 C and then transferring the culture to the storage temperature[16,19,39,43]. As mentioned earlier, the blood forms of the plasmodia and leishmania are highly resistant and in many cases withstand direct immersion into liquid nitrogen (-196 C) or a mixture of dry ice and ethanol (-76 C)[2,9,12]. The results of these experiments are quantitative rather than qualitative because protozoa protected by blood or serum (and a protective agent) will frequently survive any rate of freezing but in greatly diminished numbers[2,34,41].

The methods for achieving the controlled temperature decrease are varied. The simplest procedures are those of Eyles *et al.*[17], Cunningham *et al.*[10], and Lumsden[29]. They recommend placing the culture in a tube or vial, insulating the vial with 2-5 cm of cottoor styrene, and then placing the insulated container in a dry ice box or Revco freezer. The measured rate of cooling in these simple devices varies from 1 to 3 deg/min. Diamond[16] and others have been reported the use of a mechanical freezing unit and a liquid nitrogen vapor system for more precise control of the temperature drop.

Once a minimum temperature is reached, the final storage temperature is not important. Although most investigators[14,25,41] feel this critical storage temperature is -60 C, some find higher temperatures to be satisfactory[25,33,34]. Dalgliesh[14] thinks that the storage temperature should be at least -60 C and that best results are obtained with liquid nitrogen (-196 C).

Few investigators report successful experiments when the cultures are slowly thawed. Schneider and Seal[44] report no difference in survival of plasmodia when they are warmed at 1 deg/min or rapidly in a warm waterbath; Dalgliesh[13] obtained the same results when working with *Babesia*. Minter and Goedbloed[37] reported successful isolation of trypanosomes from tsetse flies which had been frozen in liquid nitrogen and then allowed to thaw at room temperature. Bemrick[2] obtained the best recovery of *Giardia* with a technique consisting of rapid freezing and slow thawing.

In contrast to these reports, the usual procedure is to remove the storage container from the freezing unit and plunge it into 37 C - 40 C water. Swirling the container is recommended to hasten the thawing process. In addition, most authors suggest a period of adaptation at room temperature before the culture is inoculated into media or a host. This adaptation period permits the "revitalization" of the organism before it is further stressed.

EXPERIMENTAL RESULTS

As mentioned earlier, a discussion of the results obtained with individual species is unnecessary since variation of results among species is rare. It is pertinent to mention some of the major genera and how they are affected by cryopreservation.

Plasmodia: These are unusually stable organisms which withstand most freezing techniques. In our laboratories, we have maintained five species of *Plasmodium* for over two years in liquid nitrogen. Two techniques have been used.

The preferred technique is a modification of Diamond's procedure[16]. Infective blood is drawn from a monkey, glycerol is added to a final concentration of 10%, and aliquants are added to screwcap vials. The vials are immersed in CO_2-ethanol and vigorously agitated. When the contents are completely frozen, the vials are transferred to the vapor phase of the liquid nitrogen container.

In a less frequently used procedure, no protective agent is added, but infective blood is drawn from a monkey and heparinized. Then aliquants are added to screwcap vials which are placed directly into the vapor phase of the liquid nitrogen storage container.

In either case, when inocula are needed, a vial is withdrawn and the contents thawed rapidly at 37 C; then the blood is injected without dilution into a new host. Only on rare occasions have there been no viable parasites in a vial.

Although Booden and Geiman[4] reported more quantitative results, our purpose was simply to maintain a stabilate from which new infections could be initiated. For this purpose, the technique used in our laboratory is simple and has been entirely satisfactory. It does not differ markedly from the technique used by others[22,23,42], all of whom report highly successful preservation for up to two years.

Babesia: We have maintained two unique isolates of *Babesia microti* in liquid nitrogen for up to two years. These two isolates are from human cases of babesiosis, and the retention of biologic characteristics is vital. Immediately after the cultures were established in hamsters, blood was drawn from the animals and glycerol was added to a final concentration of 10%. The blood was then frozen slowly by transferring it at hourly intervals from 5 C to -20 C to -60 C and then into the vapor phase of the liquid nitrogen storage container. Subcultures have been successful when the undiluted blood has been rapidly thawed at 37 C and inoculated into hamsters.

Both DMSO[12,38] and glycerol[10,13] have been used by various investigators in the preservation of *Babesia*. Overdulve[38] found that adding DMSO at 5% or 10% markedly increased survival of the parasite. However, since 10% gave no advantage over 5% and was somewhat toxic to the host, he recommended the use of 5% DMSO.

Trypanosoma: Using the method of Diamond[1,16], we have maintained *Trypanosoma cruzi* for over 10 years. Culture fluid was withdrawn and then the trypanosomes were centrifuged and resuspended in Locke's solution containing 10% glycerol and dispensed into glass vials which were flame sealed. The temperature was lowered 1 deg/min by slowly adding dry ice to ethanol until -20 C was reached. The vials were then placed in liquid nitrogen. After eight years the vials were transferred to the vapor phase in a new storage container where they remained for two more years. When the contents were rapidly thawed, culturing on diphasic medium was immediately successful; animals inoculated with this culture showed full virulence.

These results agree with those obtained with other techniques, but the longest time previously reported was 2 1/2 years[11]. Polge and Soltys[41] reported that after slow freezing they were able to recover 80% - 90% of the organisms "with no recognizable biological change." Herbert *et al.*[20], on the other hand, reported nearly six months' successful storage but only 20% survival. They believe this might be due to a selection of resistant organisms and suggest that investigators consider this possibility. Macadam and Herbert[30] showed by electron-micrography that the only gross injury was a dilatation of the mitochondrial envelope of the DNA core of the kinetoplast. As with the plasmodia, the trypanosomes appear to be resistant to the effects of freezing.

Trichomonas: This genus shows the best examples of species variation. *Trichomonas vaginalis* has been successfully frozen by various techniques[35,36,47]: whereas *T. foetus*, is much more fragile and requires special handling. Levine and his co-workers[24,26] obtained good survival rates when the culture was frozen in the medium in which it was grown but poor survival when the cells were washed and suspended in fresh medium. Apparently an unidentified metabolite served to protect the cells. By this method these investigators were able to maintain *T. foetus* for at least 256 days with 40% survival. With all species of trichomonads, glycerol or DMSO is required for survival.

Amoeba: All of the stock cultures of amoebae in our laboratories are preserved in the vapor phase of liquid nitrogen. Using the technique of Diamond[16] with DMSO, we have maintained cultures of Amoeba for over three years. Culture and pathogenesis studies indicate no loss of viability or virulence. It has been possible to retain cultures of differing virulence for comparisons in animal studies and antigenic analysis. Gordon *et al.*[19] reported that when they froze amoebae with DMSO, they could detect no gross cultural or morphological changes nor could they detect antigenic variations as measured by indirect hemagglutination.

Toxoplasma: The techniques for storing *T. gondii* are the same as for the other protozoa. Using the method of Eyles et al.[17], we have maintained our strains for as long as 18 months. We have also successfully used the technique of Paine and Meyer[39] in which infected tissue culture cells (human embryonic lung RU-1) were suspended in MEM with 10% DMSO and slowly frozen to -70 C. In our experience, both the rate of freezing and the rate of thawing are critical. Cultures must be slowly frozen to at least -20 C before storage and then thawed very rapidly with a minimum time lapse between thawing and inoculation into mice or tissue culture. Studies by immunofluorescence, complement fixation, and indirect hemagglutination verify that no major antigenic changes have occurred. These data agree with those of previous investigators[3,17,31,45].

Helminths: Because of their general resistance and the availability in most cases of long-lasting eggs, few investigators have attempted to freeze helminths[5,7,21,40]. In each case the attempts were only partially successful since the larvae life stage was used. Although some larvae remained viable for up to 80 days[21], over 90% did not survive, and the results were not reproducible between species. It is unlikely that this area of cryopreservation will command very much attention in the near future.

SUMMARY

It is apparent that one general method can be described which might conceivably be used for every protozoan organism. Many technical variations can be made for this one method. In some cases, such as with the plasmodia and trypanosomes, the procedure need not be followed in detail; but for others, such as with *Toxoplasma* and *Amoeba*, each step is critical. In the following generalized procedure, each step has been evaluated and used for each of the major genera we have discussed.

1. Suspend organisms (blood, tissue hemogenate, culture fluid) in a final concentration of 7.5% glycerol or 10% DMSO (in buffered saline at pH 7.2-7.4).

2. Allow 15 - 30 minutes equilibrium at 5 C.

3. Cool at 1 - 2 deg/min to at least -30 C.

4. Transfer to a permanent storage temperature below -60 C. (Take precautions to assure minimum temperature variation while samples are in storage).

5. Thaw rapidly at 37 C with agitation.

6. Allow a short (10 - 15 min) equilibration time.

7. Inoculate immediately.

REFERENCES

1. Allain, D. S. 1964. Evaluation of the Viability and Pathogenicity of Hemoflagellates after Freezing and Storage. *J. Parasit. 50:* 604-607.

2. Bemrick, W. J. 1961. Effect of Low Temperatures on Trophozoites of *Giardia muris*. *J. Parasit. 47:* 573-576.

3. Bollinger, R. O., Musallam, N. and Stulberg, C. S. 1974. Freeze Preservation of Tissue Culture Propagated *Toxoplasma gondii*. *J. Parasit. 60:* 368-369.

4. Booden, T. and Geiman, Q. M. 1973. *Plasmodium falciparum* and *P. knowlesi* Low Temperature Preservation Using Dimethylsulfoxide. *Exper. Parasit. 33:* 495-498.

5. Borrelli, D. and Trotti, G. C. 1971. Resistenza di uova di *Strongyloides papillosus* ad alcune temperature. *Parassitologia 13:* 139-143.

6. Callow, L. L. and Farrant, J. 1973. Cryopreservation of the Promastigote Form of *Leishmania tropica*. var. Major at Different Cooling Rates. *Int. J. Parasit. 3:* 77-88.

7. Campbell, W. C. and Thomas, B. M. 1973. Survival of Nematode Larvae After Freezing Over Liquid Nitrogen. *Aust. Vet. J. 49:* 110-111.

8. Coggeshall, L. T. 1939. Preservation of Viable Malaria Parasites in the Frozen State. Proc. Soc. Exper. Biol. Med. *42:* 499-501.

9. Collins, W. E. and Jeffery, G. M. 1963. The Use of Dimethyl Sulfoxide in the Low-Temperature Frozen Preservation of Experimental Malarias. *J. Parasit. 49:* 524-525.

10. Cunningham, M. P., Brown, C. G. D., Burridge, M. J. and Purnell, R. E. 1973. Cryopreservation of Infective Particles of *Theileria parva*. *Int. J. Parasit. 3:* 583-587.

11. Cunningham, M. P., Lumsden, W. H. R., and Webber, W. A. F. 1963. Preservation of Viable Trypanosomes in Lymph Tubes at Low Temperature. *Exper. Parasit. 14:* 280-284.

12. Dalgliesh, R. J. 1971. Dimethyl Sulfoxide in the Low Temperature Preservation of *Babesia bigemina*. *Res. Vet. Sci. 12:* 469-471.

13. Dalgliesh, R. J. 1972. Effects of Low Temperature Preservation and Route of Inoculation in Infectivity of *Babesia bigemina* in blood diluted with glycerol. *Res. Vet. Sci. 13:* 540-545.

14. Dalgliesh, R. J. 1972. Theoretical and Practical Aspects of Freezing Parasitic Protozoa. *Aust. Vet. J. 48:* 233-239.

15. Diamond, L. S. 1961. Freeze Preservation of Protozoa. *Cryobiology 1:* 95-103.

16. Diamond, L. S. 1961. Storage of Frozen *Entamoeba histolytica* in Liquid Nitrogen. *J. Parasit. 47:*(supplement): 28-29.

17. Eyles, D. E., Coleman, N., and Cavanaugh, D. J. 1956. Preservation of *Toxoplasma gondii* by freezing. *J. Parasit. 42:* 408-413.

18. Fulton, J. D. and Smith, A. U. 1953. Preservation of *Entamoeba histolytica* at -79 C in the Presence of Glycerol. *Ann. Trop. Med. Parasit. 47:* 240-246.

19. Gordon, R. M., Graedel, S. K. and Stucki, W. P. 1969. Cryopreservation of Viable Axenic *Entamoeba histolytica*. *J. Parasit. 55:* 1087-1088.

20. Herbert, W. J., Lumsden, W. H. R., and French, A. McK. 1968. Survival of Trypanosomes Following Rapid Cooling and Storage at -196 C. Tran. Roy. Soc. Trop. Med. *62:* 209-212.

21. Isenstein, R. S. and Herlich, H. 1972. Cryopreservation of Infective Third-Stage Larvae of *Trichostrongylus axei* and *T. colubriformis*. Proc. Helminth. Soc. Wash. *39:* 140-142.

22. Jeffery, G. M. 1957. Extended Low-Temperature Preservation of Human Malaria Parasites. *J. Parasit. 43:* 488.

23. Jeffery, G. M. and Rendtdorff, R. C. 1955. Preservation of Viable Human Malaria Sporozoites by Low-Temperature Freezing. *Exper. Parasit. 4:* 445-454.

24. Levine, N. D., Andersen, F. L., Losch, M. B., Notzold, R. A. and Mahra, K. N. 1962. Survival of *Trichomonas foetus* stored at -28 and -95 C after Freezing in the Presence of Glycerol. *J. Protozool. 9:* 347-350.

25. Levine, N. D. and Marquardt, W. C. 1954. The Effect of Glycerol as Survival of *Trichomonas foetus* at Freezing Temperatures. *J. Protozool. 1:*(Supplement): 4.

26. Levine, N. D. and Marquardt, W. E. 1955. The Effect of Glycerol and Related Compounds on Survival of *Trichomonas foetus* at Freezing Temperatures. *J. Protozool. 2:* 100-107.

27. Levine, N. D., Mizell, M. and Houlahan, D. A. 1958. Factors Affecting the Protective Action of Glycerol on *Trichomonas foetus* at Freezing Temperatures. *Exper. Parasit. 7:* 236-248.

28. Lovelock, J. E. 1963. The Mechanism of the Protective Action of Glycerol Against Haemolysis by Freezing and Thawing. *Biochem. Biophys. Acta 11:* 28.

29. Lumsden, W. H. R., Robertson, D. H. H., and McNeillage, G. J. C. 1966. Isolation, Cultivation, Low Temperature Preservation, and Infectivity Titration of *Trichomonas vaginalis*. *Br. J. Ven. Dis. 42:* 145-154.

30. Macadam, R. F. and Herbert, W. J. 1970. The Fine Structure of Trypanosomes After Preservation by Freezing. Tran. Roy. Soc. Trop. Med. Hyg. *64:* 182.

31. Mackie, M. J. 1972. Two Years Studies on the Eyles' Glycerol Preservation Technique for *Toxoplasma gondii*. *J. Parasit. 58:* 846-847.

32. Manwell, R. D. 1943. The Low-Temperature Freezing of Malaria Parasites. *Amer. J. Trop. Med. 23:* 123-131.

33. Manwell, R. D. and Edgett, R. 1943. The Relative Importance of Certain Factors in the Low-Temperature Preservation of Malaria Parasites. *Amer. J. Trop. Med. 23:* 551-557.

34. Manwell, R. D. and Jeffery, G. M. 1942. Preservation of Avian Malaria Parasites by Low-Temperature Freezing. Proc. Soc. Exper. Biol. Med. *50:* 222-224.

35. McEntegart, M. G. 1954. The Maintenance of Stock Strains of Trichomonads by Freezing. *J. Hyg.* (London) *52:* 545-550.

36. McEntegart, M. G. 1959. Prolonged Survival of *Trichomonas vaginalis* at -79 C. *Nature 183:* 270-271.

37. Minter, D. M. and Goedbloed, E. 1971. The Preservation in Liquid Nitrogen of Tsetse Flies and Phlebotomine Sandflies Naturally Infected with Trypanosomid Flagellates. Trans. Roy. Soc. Trop. Med. & Hyg. *65:* 178-181.

38. Overdulve, J. P. and Antonisse, H. W. 1970. Measurement of the Effect of Low Temperature on Protozoa by Titration. II. Titration of *Babesia rodhaini*, Using Prepatent Period and Survival Time, Before and After Storage at -76 C. *Exper. Parasit. 27:* 323-341.

39. Paine, G. D. and Meyer, R. C. 1969. *Toxoplasma gondii* Propagation in Cell Cultures and Preservation at Liquid Nitrogen Temperatures. *Cryobiology 5:* 270-272.

40. Parfitt, J. W. 1971. Deep Freeze Preservation of Nematode Larvae. *Res. Vet. Sci. 12:* 488-489.

41. Polge, C. and Soltys, M. A. 1957. Preservation of Trypanosomes in the Frozen State. Trans. Roy. Soc. Trop. Med. Hyg. *51:* 519-526.

42. Saunders, G. M. and Scott, V. 1947. Preservation of *Plasmodium vivax* by Freezing. *Science 106:* 300-301.

43. Schneider, M. D., Johnson, D. L., and Shefner, A. M. 1968. Survival Time and Retention of Antimalarial Resistance of Malarial Parasites in Repository in Liquid Nitrogen (-196 C). *Appl. Micro. 16:* 1422-1423.

44. Schneider, M. D. and Seal, N. 1973. Influence of Various Cooling and Warming Temperatures on Survival After Thawing of Cryopreserved *Plasmodium berghei*. *Cryobiology 10:* 67-77.

45. Smith, R. 1973. Method for Storing *Toxoplasma gondii* (RH strain) in Liquid Nitrogen. *Applied Micro. 26:* 1011-1012.

46. Walker, P. J. and Ashwood-Smith, M. J. 1961. Dimethyl Sulfoxide, an Alternative to Glycerol for the Low-Temperature Preservation of Trypanosomes. *Ann. Trop. Med. Parasit. 55:* 93-96.

47. Wasley, G. D. and Rayner, C. F. A. 1970. Preservation of *Trichomonas vaginalis* in Liquid Nitrogen. *B. J. Ven. Dis. 46:* 323-325.

48. Weinman, D. and McAllister, J. 1947. Prolonged Storage of Human Pathogenic Protozoa with Conservation of Virulence: Observations on the Storage of Helminths and Leptospiras. *Am. J. Hyg. 45:* 102-121.

COMMENTS ON VARIOUS ASPECTS OF THE CRYOGENIC PRESERVATION OF CELL CULTURES

A. P. Rinfret
Union Carbide Corporation

GENERAL CONSIDERATIONS.

Perusal of the foregoing papers presents convincing evidence that the technology of cultured cell banking is well developed, that the varied operations associated with such technology are carried out by institutions with widely differing objectives, and that they are performed on a routine basis. In each case a mission of basic importance to the individual institution is being served and in most instances the scientific community is the prime beneficiary. In at least one case the profit and loss statement of the corporation involved is favorably affected as the technology is effective and here the major beneficiaries are the industrial procedures of cheese and the consuming public. Institutions such as the American Type Culture Collection and the Center for Disease Control in addition to serving the day-to-day needs of biological and medical scientists for stable cell lines may also be rendering an archival function in that the present properties of living cells and viruses will be preserved intact for examination and comparison by a subsequent generation of investigators.

For the scientist planning his own frozen bank of biological material these papers will make clear that, quite apart from the ultimate objectives of which the cryogenic preservation process is a part, there is a professionalism in the procedures by which preservation is carried out, in the painstaking attention to detail required to take living cells through the changes of phase and ranges of temperature associated with long-term preservation

There are six types of operations associated with the broad function of cell preservation.

1. Selection of the population to be preserved and, if possible, optimization for preservation at a specific point in its life cycle.

2. Suspension of the cells in a suitable medium and where required, the addition of one or more cryoprotective agents.

3. The cooling operation, sometimes in a single rapid step; more often, with nucleated cells, in two or more steps under a carefully controlled regimen of temperature decline and time which will

include the phase transition from initial crystallization through whatever eutectic points happen to be present in the system.

4. Banking in the frozen state at a practicable storage temperature consistent with the maintenance of viability of the cells and their physiological and biochemical integrity for the duration of the desired storage period.

5. The thawing operation, usually carried out as rapidly as possible to maximize the yield of intact cells, but on occasion a less than maximal rate of heat transfer into the system appears to improve yield.

6. Removal or dilution of the cryoprotective reagents in the cell environment and resuspension of the cells in a medium suited to the purpose of the work.

In a practical preservation operation the various procedures to achieve success can be made routine. This, of course, does not imply that attention to detail can be neglected if there is to be maximum retention of the properties of the system. The exquisitely balanced complex which constitutes a living entity can be totally unforgiving of error in one or more steps of a given preservation process. This will be reflected in the reconstituted system in one or more ways, all of them undesirable from the point of view of the preservationist whose objective is maintenance of integrity of the system. Total loss of the capacity of the cells to reproduce themselves is catastrophic and obvious. Less obvious but, perhaps, no less catastrophic to the scientific purpose of the work would be a loss of cellular antigen, a shift in the permeability of the cellular membrane, an alteration in surface charge; more subtle changes, more difficult to detect than loss of reproductive capacity.

THE PRESERVATION LABORATORY AS A SOURCE OF NEW INFORMATION

There is no necessary relationship between the activities associated with and observations made in the preservation of cells at low temperatures and the extraction of fundamental information concerning the properties of cells as these are affected by changes in temperature, change of phase, alterations in the chemical nature of their environment, the loss or restoration of solvent or structural water, and so on. Nevertheless the cell preservation laboratory with its carefully controlled procedures is a site well fitted for the gathering of just this type of information. The preservationist by varying a single parameter in an otherwise standardized procedure on an experimental portion of the cells in his care can turn to good advantage the continuing nature of his operations to provide a steady source of new information concerning the properties of the systems with which he works.

For example, at what point in life cycle is a given organism best able to withstand the rigors of a freezing and thawing operation. Shannon states that bacteria are usually harvested at the beginning of the stationary phase; the sporulating fungi after the formation of mature spores. Trejo specifies that, in preparation of preservation, bacterial cultures should be in the active logarithmic phase. In the case of nonsporulating fungi he uses a vigorous, vegetative mycelium. These recommendations are no doubt based on experience in that they have been successful, but underlying that success are a number of biochemical, biophysical, and physiological parameters which, if known, would not only enhance the operations of preservation but would enormously increase our knowledge of cells as living systems. Would the energy status of the cell population prior to preservation correlate with viability on recovery; could the relative amounts of solvent and structural water, bound and unbound, similarly correlate? What, if any, is the effect of the cryoprotective agent on the respiratory process and is there correlation with viability, reproductive capacity, nutrient utilization or other activity following the preservation procedure?

Questions of this kind may be asked less to urge the preservationist himself to get on with such investigations than to point out to others the very fertile field for study afforded by his standardized and often extraordinarily successful laboratory operations.

THE SUSPENDING MEDIUM

The medium in which cells are suspended prior to preservation is of evident importance to their well-being not only during the periods in which cooling and freezing and warming and thawing are being carried out, but in the periods preceeding and following. Simon and Flack have indicated that the addition of 0.05M sucrose to the thawing diluent may increase the survival of *Tetrahymena vorox* and *Glaucoma chattoni*. Such observations suggest that potentially destructive but reversible conditions may exist within the cell in the immediate post-thaw period. Would the same effect have been observed were the sucrose added to the medium prior to freezing? A medium acceptable to a cell population before and during the heat removal and storage steps may be less than optimal following thawing. Were the reversible deficiency a biochemical one, it would seem amenable to remedy by appropriate modification of the pre-freeze medium. A deficiency based on some other defect, biophysical perhaps, may not yield to such an approach.

For some types of cells, particularly nucleated cells, a cryoprotective agent is an essential component of the preservation medium. A prime prerequisite for such agents is that they exercise the desired protective action at concentrations which are not toxic to the cell population. Many compounds, alone and in combination, have been investigated as protective substances in various preservation problems. Broadly, they fall into two categories; those which pass through the cell membrane and exercise their protective effect within the intra-and extracellular environments and those which do not cross the cell membrane. Glycerol and dimethyl sulfoxide (DMSO) are the best known examples of the former and polymers such as polyvinylpyrrolidone, dextran, and hydroxyethyl starch exemplify the latter. In the context of the preservation of cultured cells, the extracellular additives have not thus far played important roles.

Probably no aspect of cryopreservation has been as intensively studied as that of the concentration of glycerol or DMSO required to obtain viable cells following freezing and thawing. Any scientist seeking to preserve a cell system for which protective procedures have not been described will find such investigation indispensable to the development of a preservation protocol. The scientist interested in elaborating the basic mechanisms of cryoprotection is likely to find the interaction of protective agent, cells, and medium a rewarding area for study.

THE FREEZING PROCESS

Those involved in the operation of a frozen cell bank are thoroughly familiar with the fact that for many kinds of cells the conversion of the liquid phase to ice can be a totally destructive process as the transition takes place over a temperature range of a few degrees. With the formation of ice, the remaining unfrozen water in the system becomes ever more concentrated with respect to the solutes therein. Sodium chloride, for example, may rise to concentrations of the order of 200 grams/liter as the eutectic is approached, such concentrations being in intimate contact with the membranes of the cells compressed in channels of residual fluid lying between the crystals of ice. Buffers may be thrown out of solution in the changing concentration thus altering pH as ice formation proceeds. Intracellular water will move outward in osmotic response to the altered concentration gradient, bring subcellular structures into contact as the cell collapses, with possibly degradative interaction. The fact that otherwise vulnerable cells can withstand the rigors of the phase transition in the presence of glycerol or DMSO is evidence that over a critical range of temperature the lethal effects of solute concentration can be circumvented by slowing the conversion of solvent water to ice. Since destructive interactions which take place in an unprotected environment are chemical in nature, they will slow as the temperature declines. If they can be prevented from occurring at all until a temperature is reached at which they proceed at a negligible rate, the cultured cell population can be protected. This protective effect is accomplished by DMSO and glycerol in an extraordinary number of cases as will be evident from the communications of Shannon, Gherna, and Jung, of the American Type Culture Collection, Tejo of the Squibb Institute for Medical Research, Simon and Flack of the University of Illinois and the University of California, and Walls of the Center for Disease Control.

The rate of heat removal from a cell suspension being prepared for preservation in the frozen state can be critical to the integrity of individual cells. The relationship between heat and mass transfer during cooling and the recovery of viable, functional cells from frozen storage can be subtle and in any case is inseparable from the effects of heat and mass transfer during the warming and thawing processes. Nevertheless, it can be unequivocally demonstrated in many nucleated cell systems, suspended in specified concentrations of cryoprotective agents, that cooling can be too rapid. Cells will be killed if the rate of heat transfer is greater than that which will permit the timely migration of water from the intracellular to the extracellular environment in response to the osmotic gradient established by the formation of extracellular ice. In such situations intracellular ice, usually a lethal event, will also form. Quite possibly other phenomena are involved such as a change of phase among the lipid-containing components of the cell, or an actual dislocation at the interface of two subcellular structures due to differing coefficients of expansion. These latter, however, are largely conjecture. Their investigation could well add a new dimension of understanding to the role of heat transfer in preservation processes. Here again the cell preservationist's laboratory provides an excellent point of departure.

Nonnucleated systems such as bacteria appear to tolerate more rapid cooling rates than nucleated cells or protozoans. Sokolski, for example, freezes some of his materials by direct immersion of the vials in liquid nitrogen. Sellars emphasizes that his starter cultures for cheese production had to be frozen as rapidly as possible.

Heat transfer is, of course, a function in part of the physical system by which the rate of temperature decline is controlled. In conventional biological preparations it is also a function of the special thermal properties of water and ice. The newcomer to cryopreservation will do well to remember that the heat capacity of ice is about half that of water and that its thermal diffusivity and conductivity is greater than is the case with water. For a given system under a given temperature differential the temperature will fall more rapidly after ice has formed than before.

SPECIMEN CONTAINERS AND THE COOLING AND WARMING PROCESSES.

Most of the operations described by the authors involve cooling by moving cold gas around the vials either passively, by simple placement of the containers in the storage refrigerator, or actively, by flowing gas at a controlled rate and temperature in cooling devices designed for the purpose. Some of these devices are equipped with temperature sensing and recording apparatus which serves in the development of standardized cooling protocols and in the maintenance of processing or experimental records. Whether passive or active cooling with liquid nitrogen vapor is carried out, heat from within the cell preparation must be transferred across the walls of the container which conventionally is a relatively inefficient thermal conductor. With the exception of Sellars, who describes aluminum containers for the low temperature processing and storage of industrial starter cultures, the present authors use glass or plastic vials or ampules. Depending on the objectives of the preservationist or experimenter, it may be the thermal properties of these materials which will determine actual heat transfer rates, lowering of temperature, rather than those of the aqueous systems they contain.

The vials and ampules which predominate in cell preservation work tend to fragility at low temperatures, a hazard not found in the thermally more efficient aluminum. Nevertheless most preservationists appear to prefer glass or plastic presumably for the visibility. One area of investigation which might be rewarding to an industrial fabricator of light metals is to evaluate the thinking of laboratory personnel with respect to cell containers for frozen storage. While biological inertness is a *sine qua non* for such containers, it would be of particular interest to determine if the property of transparency or translucency outweighs such advantages as mechanical stability under changing temperature and at low temperature and such thermal properties as lower heat capacity, higher thermal conductivity and higher thermal diffusivity. Where, as is frequently the case, a rapid input of heat during the thawing process is essential such properties may have special value to the preservationist.

THE STORAGE FUNCTION.

The temperature of preservation in the frozen state for many materials of biological origin is essentially a function of the specific composition of the preparation being preserved and the objectives of the operators of the storage bank. The principal purpose of the reports of the present authors is to describe methods found effective for preservation in cryogenic systems over protracted periods of time, or, as in the case reviewed by Sellars, where absolute quality assurance in the form of viable cells capable of reproduction and enzymatic activity is essential to the success of a business operation. At a subsequent conference it is planned to discuss biological preservation procedures where the objectives of the operators of storage banks are less demanding and where higher storage temperature may be employed to the economic advantage of such banks.

The objective of most preservationists in using storage equipment operating over the cryogenic range of temperature is security. (In the biological context, the term cryogenic is applied to temperatures below -100 C.) The most commonly used equipment operating in that range is cooled with liquid nitrogen, although mechanically operated apparatus providing temperatures in its upper reaches is available commercially. The fact that refrigerators designed for liquid nitrogen use do not require compressors or other mechanical equipment, maximizes storage capacity and, for a given storage requirement, can minimize needed floor space. The absence of such mechanical equipment also eliminates the need for its maintenance and down time for its repair and replacement.

Although for many kinds of physical and industrial operations differences of this kind do not have practical significance, in applications involving biological archives, inventories of frozen cells, and distribution of potentially viable materials over geographically widespread areas (e.g., starter cultures, spermatozoa) at temperatures below -100 C refrigerators designed for liquid nitrogen use offer operating and economic advantages.

Equipment of this type is characterized by the manufacturers in terms of liquid nitrogen consumption via evaporation under conditions involving no entry or withdrawal of stored material. The result of such characterization is a volume of liquid consumed per unit time. It is not to be construed as the rate of liquid nitrogen consumption under conditions of actual use, which will vary from one application to another and from one laboratory to another. Rather, a specified evaporation rate represents the minimum amount of cryogen that will be consumed in operating the equipment whatever the actual application.

THE THAWING PROCESS.

In the context of cultured cell storage thawing is conveniently carried out by immersing the ampule in a bath and agitating same. Bath temperature, if it is desired to minimize the time required to complete the phase change, as it often is, will logically be the highest tolerated by the cells without damage. Ice being a good thermal conductor and the system being under a maximum temperature gradient, the temperature will initially rise rapidly, slowing as liquid is formed. As a practical matter it is well to remember that the heat input to liquefy ice at the melting point is almost as great as that needed to raise the temperature of a unit quantity from -196 to 0C. Agitating the ampule will promote heat transfer by convection, thus accelerating the thawing process relative to simple immersion without subsequent motion, a procedure in which the principal method of heat transfer is conduction.

The same destructive reactions that occur during the freezing process can occur during warming. When freezing in a cryogenic system, however, the temperature differential between specimen and coolant is large over the range of temperature generally regarding as critical, e.g. -0.5 to -40 C. One can use this differential, if desired, to shorten the period of time in which undesired reactions can take place. Thus gas can be flowed more rapidly around cooling vials or, by the use of appropriate techniques, vials can be treated to effect heat transfer in the nucleate boiling range at the interface with the boiling liquid nittogen, thus maximizing the rate of temperature decline. In contrast, when thawing, the temperature driving force is rapidly diminishing at just that range of temperature where it may be desirable to put heat into the system at a maximum rate. In such a situation the preservationist must take measures to promote convective heat transfer in the thawing step. Bath temperatures from 20 to 45 C can thus effect thawing in small ampules and vials in periods of about a minute or less.

THE POST THAW MEDIUM.

Following thawing, the protective additive may be removed by dilution or other procedure, pH may be adjusted, the osmotic activity of the suspending medium may be altered and so on. The point of greatest importance at this stage is that the cells may not be functioning optimally, that a period of adjustment may be necessary while they reestablish and integrate the comples of systems by which their normal metabolism is maintained.

RESPONSIBILITY IN THE STORAGE FUNCTION.

A final word on storage practice and inventory control is frozen cell banking operations. Most biological materials deemed worthy of prolonged preservation represent a heavy investment of time and money. The efforts of an investigator, perhaps extending over a period of years, to characterize the properties of a cell system and to devise preservation procedures to permit its use as a reference standard over periods of time measured in decades, or for subsequent further investigation by others in the indefinite future, can be irretrievably lost. Defective compressors, power failure, loss of vacuum in cryogenic insulation, a forgetful technician can mean the loss of the contents of a storage refrigerator, a loss which may not be recognized immediately.

Sound operation of a frozen cell bank calls for strict supervision of the storage function, the use of defensive measures (e.g. liquid nitrogen backup for mechanical systems), proper monitoring and warning systems, the assignment of responsibility to specific personnel, maintenance of an explicit inspection schedule involving not only laboratory staff but also, where available, security guards, night watchmen, and maintenance crews. In other words, sound management is essential. However, of all the precautions that the preservationist can take, none affords a greater measure of protection to his stored collection than dividing any given lot of frozen cells between two refrigerators, or, where the magnitude of the operation permits, among three or more. Placing the sole supply of a valued biological entity in a single refrigerator for prolonged storage may be quite obviously an irresponsible act.

THE LIBRARY
UNIVERSITY OF CALIFORNIA
San Francisco
(415) 476-2335

THIS BOOK IS DUE ON THE LAST DATE STAMPED BELOW

Books not returned on time are subject to fines according to the Library Lending Code. A renewal may be made on certain materials. For details consult Lending Code.

FEB 27 1990

RETURNED
FEB 14 1990

ISBN 0-309-02344-0